Geographical Information and Climatology

Geographical Information and Climatology

Edited by
Pierre Carrega

First published 2007 in France by Hermes Science/Lavoisier entitled: *Information géographique et climatologie* © LAVOISIER 2007
First published 2010 in Great Britain and the United States by ISTE Ltd and John Wiley & Sons, Inc.

ISTE Ltd
27-37 St George's Road
London SW19 4EU
UK

www.iste.co.uk

John Wiley & Sons, Inc.
111 River Street
Hoboken, NJ 07030
USA

www.wiley.com

© ISTE Ltd 2010

Library of Congress Cataloging-in-Publication Data

Information géographique et climatologie. English.
  Geographical information and climatology / edited by Pierre Carrega.
    p. cm.
  Includes bibliographical references and index.
  ISBN 978-1-84821-185-8
  1. Climatology. 2. Geographic information systems. I. Carrega, Pierre. II. Title.
  QC871.I4313 2010
  551.6--dc22

                                                                                          2009045210

British Library Cataloguing-in-Publication Data
A CIP record for this book is available from the British Library
ISBN 978-1-84821-185-8

Printed and bound in Great Britain by CPI Antony Rowe, Chippenham and Eastbourne

FSC
Mixed Sources
Product group from well-managed
forests and other controlled sources
Cert no. SGS-COC-002953
www.fsc.org
© 1996 Forest Stewardship Council

Table of Contents

Preface

Geographic information is used in many different themes and is also used as a source of information for a large number of different domains. The aim of this book is to highlight the relationship that exists between geographic information and the world of climatology. It is always a good idea to provide a definition of the subject that is being written about, so that readers do not have any misunderstandings or misinterpretations of the subject in question.

The word geography comes from the ancient Greek geo (Earth) and graphein (write). In the beginning the role of geography was to describe the Earth by creating maps. Maps are models, a way of representing what exists in reality. They are also seen as a model that can be used to transmit geographic information. Nowadays, however, the world of geography no longer only locates, observes and describes what is happening in an area. The term geography can also be applied to the study of human behavior and the environment, and whenever bio-physical areas are being studied the world of geography reminds us that these areas are very closely linked to man. Other disciplines study similar areas but what makes each of these individual disciplines different from one another is their "project", more than their actual subject content. The world of geography focuses less on the relationship that exists between man and nature, than on its spatial vision of certain phenomena. Space is to geography as time is to history, and for this reason many different studies have been carried out in areas at all levels, including studies carried out on a country, regional or territorial level, etc.

What exactly is geographical information? In order to answer this question several different responses are required so that the different chapters of this book can be understood. The term geographic can be understood here as being everything that relates to the Earth, to the interface that exists between the lithosphere, hydrosphere, and atmosphere, to the the Earth surface occupation and not only the land-use types. Every definition has its limits: do sub-soils, as well as the deepest water of the oceans form part of our study? The same logic can be applied to the air (i.e. what is not part of climatology?); does everything that exists in the air form part of our study? To avoid endless numbers of debates, which could occur on this subject, perhaps we need to adopt a certain level of pragmatism and link all of these different areas to one geographical space whenever these different areas are indirectly associated with one of the studies. In other words, if we are to understand these different areas, it is necessary to study them as closely as possible.

If geographic information takes into consideration the state of the surface of the Earth and its surrounding environment from a spatial point of view, then there is another important point that arises and needs to be dealt with: the type of geographic information that is to be produced. Nowadays, whenever the term geographic information is used it immediately involves the use of a tool known as a Geographic Information System (GIS). GISs are everywhere and are not only found in research laboratories (where they have been in use for a long time), but they can also be found in planning departments and in many administrative and local authorities. The important idea of linking one point or one pixel to a series of information with the aim of describing the point or pixel in the best possible way has been carried out by using powerful software. And the use of raster or vector GIS allows us to adapt to the different characteristics of the areas that are being studied.

It is difficult to state where the limit between the quantitative and qualitative worlds can be found in geographic terms. In this book, geographic information is more often than not the subject of the quantitative world, although this is not always the case. In the beginning, geographic information was considered in the widest terms possible and it included also the qualitative, as are often the metadata in climatology, for example. But it is true that the numbers (quantitative data) are much easier to process and deal with, as is shown in some of the different chapters of this book. Geographic information is a basis, a starting point for a series of sometimes complex operations that require multiple super positions or combinations so that a fixed goal can be reached. If quantification is compulsory each time a digital response is required, then the quantification process is also an impoverishment that is compulsory, and this is dealt with by some of the authors of the different chapters of this book: converting a measurement site into figures loses information, but what other method can be used? Several authors of this book and in particular, D. Joly, J.-P. Laborde, and P. Carrega, have been and are still faced with the following problem: as we have to digitalize information what method can we use in order to improve the process? The old issue of carrying out research on the field is raised once again. Some scientists think it is a necessary process, whilst others think that it is a time-wasting process. Although the different opinions of the different people concerned are based on strong arguments, they also depend on the individual person, for example, how they think, their memory and their mental understanding of their environment. The more operational scientists (in other words those who are committed to using concrete results) are normally those who carry out their research on site, in the field, at least in the short term.

Climatology is seen as being a domain that is capable of challenging geographic information. The field of climatology is an extremely large domain in which the number of climatologists has increased by a scale of 30 in a period of only 20 years. This book does not discuss the differences that exist between the worlds of climatology and meteorology. One major difference that does exist between these two worlds, and that should be mentioned, is the time scale that each of these domains focuses on. Meteorology focuses on forecasting what is going to happen over the short term (over a period of a few hours to a few days), whereas climatology focuses on defining, ranking, and describing events that have occurred over a longer period of time (regardless of whether this time refers to the past or to the future). The expression, "the climate was really nice today", is not used and this is due to the fact that the term climate is used to describe a relatively long period of time (at least 30 years). This means that when climate is being studied, it is possible to observe key values (average, median, etc), as well as observing the distribution of these values, and thus, the extreme values. Therefore, the difference between climatology and meteorology is more functional and temporal than it is spatial. However, the limits as to where one ends and the other begins are quite unclear. As far as the future is concerned, how do we know when the notion of climatology takes over from meteorology? This is where the notion of functionality comes into force. As far as weather forecasting is concerned, there are not very many methods that are used that can provide an accurate forecast for a period of more than 15 days. The American meteorological model known as GFS publishes a weather forecast online for up to 384 hours after the current date (in other words up to 16 days after). The European model, however, does not take as many risks and publishes a weather forecast for up to 240 hours after the current date (in other words up to 10 days after). The temporal limits of physics laws, and deterministic processes, appears when we try to predict what the weather will be like for any particular day in the future because of the non-linearity of the equations that are used in forecasting, and also because of the fact that the initial state of the atmosphere is never fully known whenever the forecasts are being calculated.

There has been an undeniable amount of progress made in the world of meteorology over the last 20 years thanks to the use of such meteorological models, and the use of other complex solutions, which P. Bessemoulin describes in Chapter 4. However, these models and solutions have spatial and temporal limits. Nevertheless, this logic (and its future updated versions), is used to forecast the average state of the atmosphere in 20, 30 or even in 100 years time. This logic is also the subject of many current debates that are taking place and which deal with the following themes: what will the climate be like in the future? What do we all need to do so that climate change can be limited? What do we all need to do in order to adapt to the changes that will inevitably take place?

Climatology is also a field that is empirical and dominated by statistics, when models, which are traditionally used in the world of physics, are unable to respond to or have difficulties in responding to the demands that exist in climatology. If a new embankment is going to be built, working out its height involves considering the water levels that were measured in the area in the past. If these measurements are adjusted by Gümbel's distribution (for example), it then becomes possible to work out the probability that a certain level of water will be exceeded, and thus its "return period". All of this information should form part of what is known as a stationarity hypothesis, which nowadays is not normally validated.

Bringing together models from both the worlds of physics and statistics is a usefull exercise from an an intellectual point of view, although formal, because interactions between models from these two domains occur on a daily basis. Each physical model relies ever so slightly on the use of calibration coefficients that are determined by statistics, and inversely, each effective and operational statistical model that is used in climatology relies on the use of different fundamentals that stem from the world of physics.

The most common methods that are used today include: multiple regressions, and geostatistics based on spatial autocorrelation (kriging in particular), which are sometimes combined. The use of neural networks is not as widespread, and this method does not seem to solve many of the issues that people thought it would be able to a few years ago.

Remote sensing is a term that is used to group together all of the different tools that are able to record information from a distance, which is usually done using airplanes or satellites. The multiple sensors, which can be found on board these vessels, can contribute to collecting geographic information that can be used to recreate the relief of an area or to evaluate how well a particular crop is growing from a phenological point of view, etc. Sensors can also be used to measure different climatological variables, such as the temperature of the Earth's surface or the temperature of the clouds as is explained in Chapter 3 by Dubreuil. What makes remote sensing different from other methods is that it can provide data on two different pieces of information that are being researched at the same time, or at least in part.

There is one fundamental issue that affects geographic information and the relationship that it has with the world of climatology: how is it possible to make these two different domains evolve together in the future? Roussel, the author of Chapter - 6, reminds us that the geographic information produced depends on the metrological and political context in which it is used. With this in mind, different rules and regulations, as well as different socio-economic contexts and the mentality of the general public, will influence how the geographic information is used. Advances in technology in the future will probably change the way in which geographic information is measured, and as a result what is actually being measured. Will financial fluxes be a more important part of geographical information in the future?

This book is made up of eight chapters, and can be divided into two main parts.

The first part of the book is devoted to the technical aspect and the tools used to gather geographic information. In Chapter 1, Wolfgang Schoner analyses the bases of climatological observations for GIS applications, while in Chapter 2 Daniel Joly focuses on spatial analysis and cartography, and throughout the chapter he elaborates on the use of the statistical approach. In Chapter 3, Vincent Dubreuil shows how remote sensing can be used to provide us with both geographic information and information relating to the climate. In Chapter 4, Pierre

Bessemoulin provides us with an explanation of a number of key elements that are used to give us a better understanding of the way in which meteorological and climate models exploit geographic information so that they can be used effectively.

The second part of the book is devoted to how the geographic information is applied to different domains. The themes that we have chosen to focus on are associated with risks or certain constraints. The characteristics of the climate as it is today are associated with the actions of man, his needs and his limits. The research that we have carried out focuses on these limits. In Chapter 5, Maria Joao Alcoforado shows the necessity of geographical information to understand the specificity of urban climates; and, in Chapter 6 Isabelle Roussel focuses on the complexity of the relationship that exists between climatology, atmospheric pollution, and geographic information. Throughout the chapter she shares her views on what the term geographic information means and in some cases questions the term itself.

In Chapter 7, Jean-Pierre Laborde, who is a passionate hydrologist and renowned technician, proves that it is necessary to take a step back to understand exactly what a simple water flow or fload means. By taking spatialized geographic information into consideration he places a lot of importance on climatology. Finally, in Chapter 8, Pierre Carrega defines meteorological risk levels associated with forest fires. He bases his research on two different methods that can be used to generate the meteorological risk level index and compares them throughout the chapter. The two methods are both part of geographic information and the world of climatology.

Pierre CARREGA

Chapter 1

Basics of Climatological and Meteorological Observations for GIS Applications

Weather and climate data are spatially distributed. Geographical information technologies can therefore provide a useful and relevant working environment for the distribution, integration, visualization, and analysis of these data. However, compared to other scientific areas, the application of geographical information system (GIS) tools was for a long time a clumsy process within meteorology and climatology, and especially within most national meteorological services (NMS); because of the shortcomings of GIS related to the underlying data model and missing interfaces to standard meteorological tools (e.g. weather forecast model). While the GIS data models are highly static based, meteorological data models have a need for a strong dynamical component with causal dependencies in the space/time domain (see for example [CHR 02]). Nativi *et al.* [NAT 04] describe the differences between both underlying data models and advocate models that are supported by so-called interoperability services. In addition to these differences in the data models, there are significant differences in the spatial modeling approaches. In general, GIS environments have implemented the geo-statistical modeling tools that are based on one temporal realization only, whereas meteorological data offer the temporal sample in addition to the spatial sample, which results in different spatial modeling approaches [SZE 04]. However, within the last few years efforts for integration of meteorological data models in GI environments were quite successful, and well-established GIS web-mapping standards and spatial infrastructures have gained increasing importance in meteorology and climatology. Thus parallel efforts and development currently appear to be resolved [SHI 05].

Information to be derived from climate variability analyses is strongly dependent not only on the spatiotemporal density, but also on the quality of the available data. Today it is a well-established fact in climatology that the climate signal from measurements, beside the statistical noise, is by inhomogenities. Therefore, a primary step of climate studies is to analyze the input data used with respect to

---

Chapter written by Wolfgang SCHOENER.

quality and homogenity, which makes the results uncertain regarding input data, in addition to the uncertainty of the approach. Quantifying quality and homogenity of the data require data about the data itself (the metadata). Nowadays, metadata are highly standardized for GI data (e.g. see the Open GIS Consortium activities), but the information obtained from metadata regarding climate data is still heterogenous. However, the NMSs and the WMO (World Meteorological Organization) are aware of the importance of climate metadata, which resulted in several efforts for standardization of metadata (see e.g. [AGU 03]). To summarize these efforts, it can be stated that in climatology metadata information is related to the documentation of the "where", "when", and "how" of measurements, whereas metadata in GI science also add emphasis to the usability of the data.

In this chapter, basic concepts of climate networks and climate data are presented. This includes an overview of standards of climate measurements, description of climate data types, spatial reference of data, as well general comments on accessing the data. The areas of climate data quality and homogenity are reviewed in depth, covering the important aspects of metadata description. The chapter does not tackle climate model data and only introduces climate reanalysis data.

1.1. Data measurements and observations in climatology

1.1.1. *Networks and concepts for meteorological/climate data*

Meteorological measurements are motivated by the primary aim of predicting the Earth's weather with the highest possible precision. This aim results in measurements covering the entire Earth (for both the land and the sea), but also encompassing the third dimension (vertical sounding of the atmosphere by radio sounds, satellite sensors, radar, etc). Beside weather forecasting, meteorological services are responsible for monitoring the state and spatiotemporal change of the climate. As these two basic aims do not coincide with respect to network performance, two different networks have been established in public weather services, the synoptic and the climate network. Whereas, the stations and instruments are identical, the networks differ in their interval, quantity, availability and time of observations. Moreover, the synoptic network is characterized by the need for a much larger spatial extent and more detailed information on past and current weather situations. In contrast, climate networks are characterized by higher demands on data quality. All national meteorological/climatological networks are coordinated on an international level by the WMO.

The need for meteorological/climatological networks is met by *in situ* measurements and by remote sensing techniques. Consequently, the WMO Global Observing System is composed of the surface-based subsystem and the space-based subsystem. The surface-based subsystem includes different types of station networks (e.g. surface synoptic stations, climatological stations), whereas the space-based subsystem comprises, for example, on-board sounding from spacecraft. The observational requirements of a climatological station or synoptical station are detailed in [WMO 03] and include: present weather, past weather, wind direction and speed, cloud amount, cloud type, cloud-base height, visibility, air temperature, relative humidity, air pressure, precipitation, snow cover, sunshine duration or solar radiation, soil temperature, evaporation.

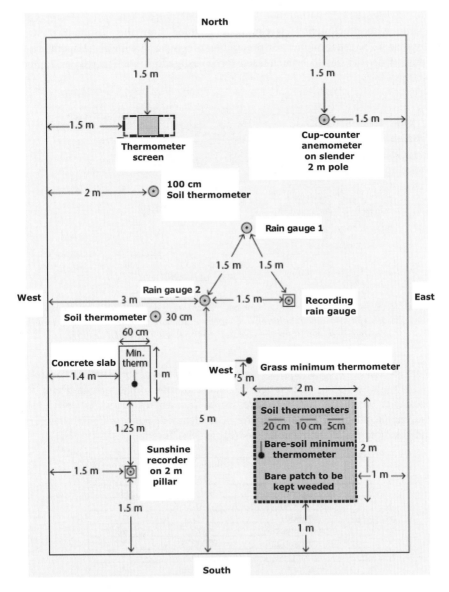

Figure 1.1. *Layout of an observing station in the northern hemisphere showing minimum distances between installations (from [WMO 08])*

An important concept behind climate observations is representativeness, which is the degree to which the observation accurately describes the value of the variable needed for a specific purpose. Therefore, it is not a fixed quality of any observation, but results from joint appraisal of instrumentation, measurement interval and exposure against the requirements of some particular application [WMO 08]. An estimate of spatiotemporal representativeness of air temperature and precipitation is shown in Figure 1.2, with much higher spatial correlation for air temperature compared with precipitation. It can be concluded from this results that station density has to be much higher for precipitation compared with air temperature and that station density has to be increased for investigations with increasing temporal resolution.

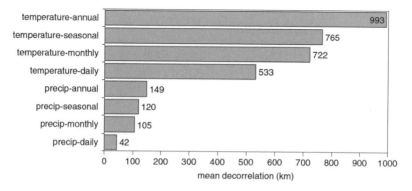

Figure 1.2. *Average decorrelation distances (r² decreasing below 0.5) for air temperature and precipitation in four time resolutions. Samples: daily values for all of Europe; monthly, seasonal and annual for the Greater Alpine Region (from [AUE 05])*

Various meteorological applications have their own preferred timescale and space scale for averaging, station density, and resolution of phenomena. From there, for example, weather forecast requires more frequent observations compared to climate monitoring. The spatio-temporal dependency of meteorological phenomena results in simple scaling convention (see Table 1.1).

Type of motion	Spatial scale (m)	Temporal scale
Eddy	0.001	0.001
Micro turbulence	10	10
Tornado	100	60
Cumulus convection	1,000	20 min
Cumulunimbus	100,000	1 h
Front	100,000	3 h
Hurrican	100,000	3 h
Cyclone	1,000,000	1 d
Planetaric waves	10,000,000	10 d

Table 1.1. *Spatial and temporal scales of meteorological phenomena*

The design of a meteorological station has to be according to the network requirements. In particular, the station site, instrument exposure and location of sensors has to be treated according to regulations. As an example, Figure 1.1 shows the layout for a typical synoptic/climatological station according to WMO regulations.

1.1.2. *Standards for climate data measurements*

The term "standard" is related to the various instruments, methods, and scales used to estimate the uncertainty of measurements. Amongst others, nomenclature for standards of measurements is given in the *International Vocabulary of Basic and General Terms in Metrology* issued by the International Organization for Standardization (ISO) [ISO 93]. The following standards are included: measurement standard, international standard, national standard, working standard, transfer standard, traceability, etc.

Meteorological observations and measurements are highly standardized from WMO or from NMSs. Such standardization is obvious if we takes into account the influence of station surroundings (surface properties, influences from nearby buildings, trees, etc) or of measurement observation procedures. From there meteorological (climatological) measurements are standardized especially with respect to:

– surface conditions in the nearby of the sensor;

– station surrounding;

– sensor-height above ground;

– procedure of reading;

– observation time.

However, practices are different and measurements are occasionally performed under conditions that are different from the required standard, which have to be archived in metadata information. This is especially true for the surface conditions in the areas around the sensor and the station, whereas sensor height and observation are generally in accordance with the standards. Standards are more accurately considered in climate networks of weather services compared with networks from other operators. When incorporating data from various other sources, the standardization regulations of the data providers should be carefully considered.

1.1.3. *Climate data types*

Classification of data types can be undertaken from different perspectives. Using classical classification schemes used in GI science the following types of data are used in meteorology and climatology.

– *Spatial irregularly distributed point data*: e.g. the station measurements and observations, vertical radio sounding data if some generalization is taken into account;

– *Raster data*: e.g. the different field from weather forecast models or from climate models;

– *Image date*: e.g. satellite data, weather radar data.

Meteorological data can also be classified into scalar data (air temperature) and vector data (wind with wind speed and wind direction). According to the classical basic of statistics meteorology/climatology include all types of scales of measurements:

– *Nominal scale*: e.g. cloud type, present weather, weather type;

– *Ordinal scale*: e.g. cloud density;

– *Interval scale*: e.g. air temperature;

– *Ratio scale*: e.g. precipitation, air pressure.

In addition to these statistical or GIS-related classification schemes, there are also such from meteorology/climatology schemes based on the idea of a Global Observing System [WMO 08]:

– Surface-based subsystem: comprises a wide variety of types of stations according to the particular application (e.g. surface synoptic stations, upper-air stations, climate stations);

– Spaced-based subsystem: comprises a number of spacecraft with on-board sounding missions and the associated ground segment for command, control and data reception.

1.1.4. *Access to climate data: spatial data infrastructure in meteorology and climatology*

Since the foundation of the NMS, the weather forecast has been highly dependent on efficient spatial data infrastructure, which today is called the Global Telecommunication System (GTS) and covers the entire Earth. Station observations and other data are shared with GTS worldwide within the hour according to standardized regulations, in order to get a "snapshot" of the current state of atmosphere and weather conditions and as an input for weather forecast models. The GTS data infrastructure is highly standardized and secures the data transfer between the NMSs, but does not fully meet the needs of the increasing number of users outside the NMSs' networks. As a result, international NMSs networks, such as EUMETNET (The Network of European Meteorological Services), established projects to address this, e.g. UNIDART (uniform data request interface, http://www.dwd.de/UNIDART). During the last few years, and based on GTS, WMO initiated the WIS (WMO Information System), which distributes information globally for real-time weather forecasting and climate monitoring using a service-oriented architecture.

The operability of the GTS is dependent on data exchange and a related policy of data holders. Today, each National Meterological and Hydrological Service (NHMS) has its own data access policy ranging from free access to highly commercially oriented data selling. Even within the European Union (EU), meteorological data policy is quite heterogenous and the exchange of data between NMHS, apart from for weather forecast purposes, such as for climate monitoring, is sometimes limited. Generally, important information on meteorological data is provided in table or map form by basic metadata of the station network, which provides information about location, geographical coordinates, altitude, sensor equipment and data availability, etc., without any charge from NMHS. Easy access to such information is still not guaranteed, but there is a move towards providing a greater amount of information without restriction.

In Europe, the idea of spatial data infrastructure (SDI) was substantially supported by the INSPIRE (http://inspire.jrc.it/home.html) initiative. INSPIRE is an EU directive that forces EU member states to provide spatial data to different users according to OGCs SDI standards. As a result of INSPIRE and as a general need of climate research, European NMSs started with efforts to meet INSPIRE needs. Within the frame of EUMETNET, the EUROGRID was formulated with its first step as a showcase (S-EUROGRID, see www.eurogrid.eu, [KLE 08]). EUROGRID aims to provide a SDI for climate data according the OGC standards. In addition to this multinational initiative, climatological/meteorological SDIs were established on national levels. SeNorge, a common meteorological and hydrological effort in Norway, is a good example (www.senorge.no). Due to the user-friendly data policy in Norway, SeNorge not only displays climate data fields on a monitor screen according to the OGS WMS (Web Map Service) standard, but users can also obtain and integrate data of interest according to the WFS (Web Feature Service) and WCS (Web Coverage Service) standard. These OGC standards for web-mapping have received substantial interest in the field of meteorology over the last few years.

Another well-established OGC standard used in addition to meteorological and climatological applications is the Google Earth KML format for many web services. Integration of OGC-compliant spatial infrastructure for distribution of climate data received much earlier support in the USA compared with Europe. The NOAA (National Oceanic and Atmospheric Administration), and in particular NCAR (National Centre for Atmospheric Research), supported the OGC ideas of interoperability for meteorological data. Special attention was given to the ArcGIS Atmospheric Data Model, a collaborative initiative among ESRI, UCAR, NCAR, Raytheon, Unidata, and NOAA. The ArcGIS Atmospheric Data Model aims to represent each of these data objects in a uniform manner, enabling their superposition and combined analysis in the ArcGIS desktop environment. For the first time, the ArcGIS 9.2 [ESR 09] release supported both the NetCDF and HDF-5 data format through a new tool from the ArcGIS toolbox list. Both the NetCDF and HDF data models are commonly used in atmospheric sciences, e.g. data fields from climate model runs are available in NetCDF. Through this data model, a fundamental linkage between the GI community and atmospheric sciences community was established.

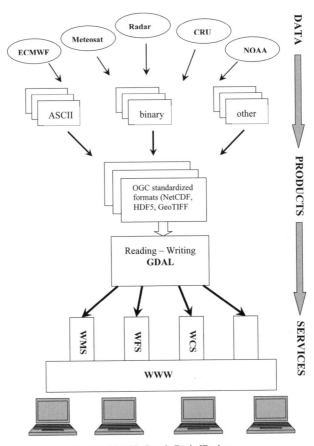

Figure 1.3. *Simplified scheme of OGC compliant web services (GDAL stands for Geospatial Data Abstraction Library, figure adapted from [VAN 08])*

In particular, the GALEON IE (Geo-interface for Atmosphere, Land, Earth and Ocean netCDF) Interoperability Experiment supports open access to atmospheric and oceanographic modeling and simulation outputs. The geo-interface to netCDF datasets is established by the Web Coverage Server (WCS 1.0) protocol specifications. Additionally, UNIDATA unified the OpenDAP, netCDF and HDF5 data models to the new CDM (Common Data Model) and introduced a new API (application programming interface), NcML, an XML (extensible mark-up language) representation of netCDF using XML syntax. On a long-term perspective, GALEON will analyze FES (Fluid Earth Sciences) requirements for simple and effective interface specifications to access datasets and will define a more general data model for CF-netCDF. This new data model should include non-regular data grids and should establish metadata encodings (e.g. Climate Service Modeling Language CSML, ncML-G). CSML is a standard-based data model described in Unified Modeling Language (UML), and an XML mark-up language that implements this data model [WOO 06]. The model describes climate science data (e.g. observational data, model runs) at the level of the actual data values; CSML is not a high-level discovery metadata model [LOW 09]. An example of a simplified structure of an OGC compliant web service for integration of meteorological/climatological data in geospatial services or applications is shown in Figure 1.3.

Another major OGC initiative with increasing interest from meteorology is Sensor Web Enablement (SWE). The ultimate goal of SWE is to make all kinds of sensors discoverable, accessible, and controllable via the web, which should result in "plug-and-play" web-based sensor networks. Beside others, SWE include Sensor Observation Service (SOS), Sensor Planning Service (SPS), and Sensor Alert Service (SAS). SOS aims to provide access to observations from sensors in a standardized way that is consistent for all sensor systems, including remote, *in situ*, fixed and mobile sensors.

As mentioned previously, the time dimension is an important domain in meteorology and climatology not adequately covered by GIS (see e.g. [WOO 05]). Moreover, climatology and weather forecast are highly interested in slices of time, showing climate fields on axes of latitude and time or longitude and time. Such diagnostic slices are required in future GIS standards. In addition to the time dimension, the representation of gridded meteorological fields could result in problems. For instance, meteorological grids can be non-regularly spaced or, in the case of models formulated in spectral coordinates, could have fewer longitudinal grid-points towards the poles. These shortcomings need to be addressed in the future by additional cooperating standardization work between GIS and meteorology.

1.1.5. *Spatial reference for climate data*

The position of climatological/synoptic station has to be measured in the World Geodetic System 1984 (WGS-84) or Earth Geodetic Model 1996 (EGM96). The coordinates of a station includes [WMO 08]:

a) the latitude in degrees with a resolution of 1 in 1,000;

b) the longitude in degrees with resolution of 1 in 1,000;

c) the altitude of the station above mean sea level to the nearest meter.

The elevation of the station is defined as the altitude above mean sea level of the ground on which the rain gauge stands or, if there is no rain gage, the ground beneath the thermometer screen. If there is neither a rain gauge nor screen, it is the average level of terrain in the vicinity of the station. If the station reports air pressure, the elevation of air pressure sensor must be specified separately.

Within the last few years, the increasing number of spatial modeling tools with increasing spatial resolution used in meteorology and climatology also enforced the pressure on the accuracy of station coordinates. Previously station coordinates were digitized from topographical maps; currently, station coordinates are surveyed by

GPS measurements. In addition to the spatial reference in geographical coordinates according to WMO, there are still a great number of national reference systems in use. However, the increasing use of the UTM system will overcome this variety of spatial reference systems in the future. Problems from different national reference systems could appear in the case of merging datasets (especially gridded fields) from different data holders, which could result in certain differences in overlapping areas.

In addition to altitude, several other vertical coordinate systems are used in meteorology including pressure, isentropic, or terrain-following coordinates, which are used for upper-air observations or for weather and climate models. Beside upper-air observations, such coordinated systems are used for weather and climate models. Such systems are not established in the traditionally predominately two-dimensional GIS world [WOO 05]. Providing the full richness of vertical coordinate systems will be an important requirement for full integration of GIS in meteorology and climatology, and thus, an important area of OGC activity.

1.1.6. *Climate reanalysis data*

Climate reanalysis aims to produce meteorologically consistent datasets of the atmosphere covering the entire Earth with state-of-the-art methods. In particular, they combine the full set of meteorological observations, including, e.g. surface stations, radio sounds, and satellite data with weather forecast models using data assimilation methods. Climate reanalysis datasets are among the most important climate datasets in climate research, including climate impact studies. Standard data formats for climate reanalysis are GRIB or NetCDF. Due to the NetCDF data format, these datasets are already standardized for direct use in GIS applications (see section 1.1.4). Climate reanalysis data are provided in Europe by the ECMWF (European Center for Medium-range Weather Forecast, UK) from the following projects: ERA15 covering the period 1979-1993 and ERA40 covering the period 1957-2002. In the USA, reanalysis projects have been run by the NOAA and NASA within: NOAA-NCEP covering the period 1948 onwards and NASA/DAO covering the period 1980-1995, and from Japan Meteorological Agency: JRA-25 covering the period 1979 onwards.

New reanalysis projects are currently under way (ERA interim) or planned (NCEP, JRA). In addition to meteorological consistency, the most important product of reanalysis data is their full spectrum of data covering the entire atmosphere in similar way as weather forecast models in high temporal resolution (e.g. 6 hourly fields for ERA40), but also with similar spatial resolution of the gridded fields.

Climate reanalysis is derived by data assimilation methods, which is today a four-dimensional (4D) variational analysis in the case of ERA [AND 08]. 4D-Var performs a statistical interpolation in space and time between a distribution of meteorological observations and an *a priori* estimate of the model state (the background). This is done in such a way that the dynamics and physics of the forecast model is taken into account to ensure the observations are used in a meteorologically consistent way. The idea behind 4D-Var data assimilation is shown in Figure 1.4. For a single parameter x the observations are compared with the short-range forecast from a previous analysis over a 12-hour period. The model state at the initial time is then modified to achieve a statistically good compromise, x_a, between the fit, J_b, to the previous forecast, x_b, and the fit J_0 to all observations within the assimilation window. J_b and J_0 are referred to as cost functions [AND 08]. This 4D-Var approach replaced earlier approaches that were based on the optimum interpolation method.

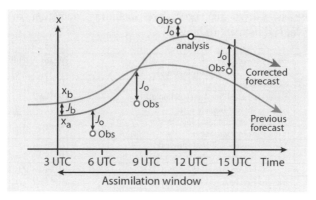

Figure 1.4. *The idea of 4D-Var data assimilation technique, see the text for a detailed explanation (from [AND 08])*

Climate re-analyses are also subject to a detailed validation against independent observations. A detailed description of climate reanalysis goes far beyond the scope of this chapter; for more details the interested reader should refer to the literature (e.g. [UPP 04]).

1.1.7. *Climate data providers outside NMHs*

Climate data are not only provided by meteorological and hydrological services but also by other data providers, in particular, universities. The majority of these data centers provide surface climate data that are either station data or gridded data. Usually, these data providers use data from NMHs and improve their data quality or spatial coverage. From the GIS perspective, standardization of these datasets is still weak and both data formats and metadata are quite heterogenous. Consequently, the import of the data to GIS needs some data preparation. However, OGC standardization of the NetCDF format is expected to overcome this shortcoming in the near future.

A major climate data provider is the Climate Research Unit (CRU) from the University of East Anglia (UK). In particular, CRU provides long-term climate data with global coverage, which is an important base for global climate monitoring.

12 GIS and Climatology

1.2. Data quality control and data homogenization in climatology

1.2.1. *The importance of data quality control and homogenization*

Data quality control (DQC) is applied to detect errors in the process of recording, manipulating, formatting, transmitting and archiving data. DQC is not identical to homogenization as homogenization goes far beyond the aims of DQC. For homogenization, long-term series of climate data are needed, which enable non-climatic breaks to be detected in the series resulting from changes of station location, observer, observation time, sensor type, station surrounding, etc. In fact, homogenization is a two-step procedure including detection of breaks with statistical tests and adjustment of breaks.

Country	Data provider		Means calculus	Time
Austria	Central Inst. for Meteorology and Geodynamics	ZAMG	$(t_1 + t_\mu + 2*t_p)/4$	LMT
	Hydrographical Service (yearbooks)	HZB	$(t_1 + t_n)/2$	LMT
Bosnia and Hertzegovina	Federal Meteorological Inst.	Meteo BiH	$(t_1 + t_{14} + 2*t_{21})/4$	LMT
	Federal Meteorological Inst. (historic Yugoslavian yearbooks)		$(t_1 + t_{14} + 2*t_{21})/4$	
Croatia	Meteorological and Hydrological Service of Croatia	DHMZ	$(t_1 + t_{14} + 2*t_{21})/4$	LMT
Czech Republic	Czech Hydrometeorological Inst.	CHMI	$(t_1 + t_{14} + 2*t_{21})/4$	LMT
France	Météro-France		$(t_1 + t_n)/2$	-
Germany	German Meteorological Service	DWD	$(t_1 + t_{14} + 2*t_{21})/4$	1961-86: LMT 1987-90: CET +30'
Hungary	Hungarian Meteorological Service	OMSZ	$(t_1 + t_{14} + 2*t_{21})/4$	LMT
Italy	Italian National Research Council, Inst. of Atmospheric Sciences and Climate	ISAC-CNR	$(t_1 + t_n)/2$	-
	University of Milan, Dept. of Physics	UNIMI	$(t_1 + t_n)/2$	-
	University of Padua, Treeline Ecology Research Unit	UNIPD	$(t_1 + t_n)/2$	-
	University of Pavia, Dept. of Territorial Ecology and Terrestrial Environments	UNIPV	$(t_1 + t_n)/2$	-
	University of Turin, Dept. of Agronomy, Forest and Land Management	UNITO	$(t_1 + t_n)/2$	-
	Giancarlo Rossi, private data collection		$(t_1 + t_n)/2$	-
	Italian Meteorological Society, Aosta Valley, Piedmont	SMI	$(t_1 + t_n)/2$	-
	University of Turin, Department of Earth Science, Piedmont	UNITO	$(t_1 + t_n)/2$	-
Slovenia	Environmental Agency of the Republic of Slovenia, Climatological Dept.	ARSO	$(t_1 + t_{14} + 2*t_{21})/4$	LMT
Slovakia	Slovak Hydrometeorological Inst.	SHMU	$(t_1 + t_{14} + 2*t_{21})/4$	LMT
Switzerland	Federal Office of Meteorology and Climatology	Meteo Swiss	$(t_1 + t_{14} + 2*t_p)/4$	CET +30'

Table 1.2. *Example of the heterogenity of air temperature station networks for the Greater Alpine Region GAR used for spatial modeling of climate normal fields 1961-90 (tn=mean daily minimum temperature; tx=mean daily maximum temperature; TRM=true mean; CET=Central European Time; LMT=local mean time, adapted from [HIE 09])*

As a minimum requirement, a yes/no answer is recommended to indicate whether DQC has been applied or not. If the answer is positive, it would be good practice to describe the degree of DQC applied to the data (e.g. subjected to logical filters only; compared for internal coherency in sequence of observations, for spatial consistency among suitable neighboring stations, for coherency with its climatological values and limits) and to provide details on the employed techniques and their application [AGU 03].

A simple example of inhomogenity in climate data series results from different approaches for the computation of daily or monthly means of e.g. air temperature from either observations at fixed times or from daily extremes of temperature (Table 1.1). The example shown in Table 1.2 is taken from the work of a new air temperature map for the Greater Alpine Region (GAR) [HIE 09] using powerful spatial modeling approaches including GIS techniques. However, before spatial modeling could be started, station measurements had to be transformed to common mean formula. Beside the formula for mean computation, the time reference system used is also heterogenous within the GAR study region. It is obvious from this simple example (which only tackles one out of several inhomogenities in climate datasets) that DQC and data homogenization, in particular, are a laborious part of climate modeling studies. Exclusion of this part of the modeling study could result in systematic biases.

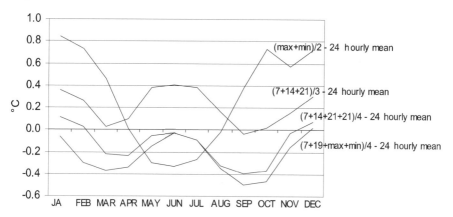

Figure 1.5. *Evolution through the year of the difference between various ways of calculating daily mean temperature and 24-hourly observations average for the inner-alpine station Puchberg in Austria, 1987-1996. Data source: Central Institute for Meteorology and Geodynamics, Vienna, Austria (from [AGU 03])*

Long-term series from measurements of automatic weather stations with hourly values make it easy to compute the differences between various computation formulas of daily means of air temperature used by NHMs. Selected examples of differences between commonly used mean formulas and a 24-hourly mean are shown in Figure 1.5. In fact, the widely used formula of (max+min)/2 show differences of up to 1°C to the 24-hourly mean, which turns out to be larger than the final standard error of spatial modeling of air temperature in the case of the GAR study example. Similarly, it was shown by many studies that inhomogenities can even exceed the climate change signal in climate time series (see e.g. [AUE 07]). It is quite easy to understand from these findings that treatment of data homogenity is essential in the analysis of spatial or temporal variability in climate data.

Although adjustment of errors originating from different means calculations can be performed quite easily in the case of longer time series from automatic weather stations, adjustment of inhomogenity originating from e.g. urbanization effects of villages is not that simple. Urbanization does not cause a sudden break in series but instead a gradual inhomogenity trend (Figure 1.6). In the case of homogenization of the urbanization effect, it is very useful to collect information on changing building density and changing land-use.

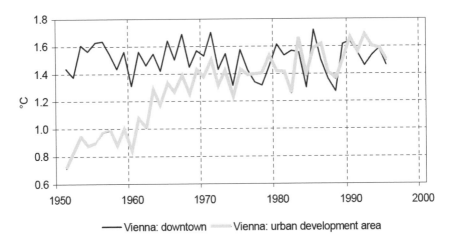

Figure 1.6. *Time series of annual mean urban temperature excess (relative to rural mean 1951 to 1995) based on height-reduced temperature records. The station in the densely built-up area shows a stable temperature excess against the rural surroundings, whereas the trend of temperature excess at the station in the urban development area is 0.18°C per decade. Data source [BOH 98]*

Another inhomogenity in climate networks results from the change of sensors of the same type or different types. The increasing number of automatic weather stations causes such a systematic shift of sensors. Parallel measurements with both the old and the new sensor correctly merge the datasets of different sensors. However, such parallel measurements are not performed on a regular basis, and even if parallel measurements are undertaken, they are quite often undertaken over a very short period. An example of inhomogenity from different sensors is shown in Figure 1.7 for measurement of sunshine duration in Austria, replacing the Campbell-Stokes sunshine autograph with the Haenni-Solar sensor.

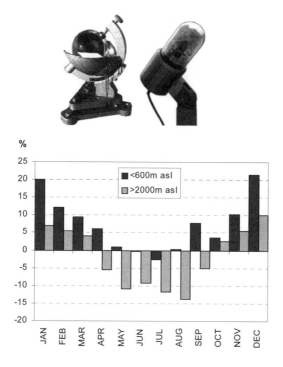

Figure 1.7. *Top: two types of instruments to record sunshine duration used in the Austrian meteorological network: Campbell-Stokes sunshine autograph and Haenni Solar system of automatic weather stations. Down: Consequences: Mean annual course of the breaks in Austrian sunshine series due a change from the traditional Campbell-Stokes recorders to the Haenni-Solar sensors of the automatic network (new minus old in %, sample 1986-1999, dark: mean of four low-level sites, light: mean of three high-level sites) [AUE 01]*

	air pressure	temperature	precipitation	sunshine	cloudiness	all	
no. of series	72	131	192	55	66	516	series
available data	10215	19312	26063	7886	7669	71145	years
mean length of series	141.9	147.4	135.7	88.8	119.5	137.9	years
detected breaks	256	711	966	366	234	2533	breaks
mean homogeneous sub-interval	31.1	22.9	22.7	11.6	26.3	23.4	years
detected real outliers	638	4175	529				outliers
filled gaps	4217	12392	14927	2011	3513	37060	months
mean gap rate	3.4	5.3	4.8	2.1	3.8	4.3	%

Table 1.3. *Result of a homogenity study from monthly multiple climate series from the Greater Alpine Region (from [AUE 07])*

A very detailed study on climate data homogenity is available for the Greater Alpine region from several research projects (e.g. [AUE 07]). Some important results of these studies are summarized in Table 1.3 showing that data inhomogenity is immanent to climate data studies even on a very short time scale. The number of detected breaks is quite high and the mean length of homogenous sub-interval is in the range of 10-30 years for all climate variables shown. Results in this table are derived from selected monthly data series covering monthly means and monthly sums. If, however, climate extremes or daily data series are studied, the problem of data homogenity is even more pronounced.

1.2.2. *Methods for climate DQC*

DQC is part of the core of the whole data-flow process. In fact, it has to ensure that data are checked and is as error-free as possible. All erroneous data have to be eliminated and, if possible, should be replaced by corrected values (while retaining the original values in the database).

Useful tools of DQC for climate data are (Aguilar *et al.*, 2003):

a) *Gross error checking*: report what kind of logical filters have been utilized to detect and flag obviously erroneous values (e.g. anomalous values, shift in commas, negative precipitation, etc).

b) *Tolerance test*: documents to which tests have been applied, to flag those values considered as outliers with respect to their own climate-defined upper/lower limits. The tests provide the percentage of values flagged and the information on the approximate climate limits established for each inspected element.

c) *Internal consistency check*: indicate whether data have undergone inspection for coherency between associated elements within each record (e.g. maximum temperature < minimum temperature; or psychrometric measurements, dry-bulb temperature ≤ wet-bulb temperature).

d) *Temporal coherency*: inform if any test has been performed to detect whether the observed values are consistent with the amount of change that might be expected in an element in any time interval and to assess the sign shift from one observation to the next.

e) *Spatial coherency*: notify if any test is used to determine whether every observation is consistent with those taken at the same time in neighboring stations affected by similar climatic influences.

Figure 1.8 shows the results from a detailed homogenization study of climate time series for the GAR, which also included estimation of outliers and gap filling. Whereas the time series of outlier rates (figures on the left) indicate more about internal system stability of meteorological networks the gap rates (figures on the right) seem to react more to external influences. It is interesting to see from Figure 1.8 that both outliers and gaps increased since the 1980s, which was the beginning of automation of climate networks in the study region.

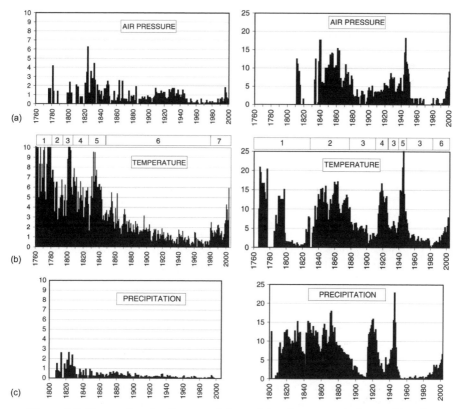

Figure 1.8. *Time series of outlier rates (left panels) and gap rates (right panels) in the HISTALP series of (a) air pressure, (b) air temperature, and (c) precipitation. Outlier and gap rates in percentage (in relation to the amount of available data) (from [AUE 07])*

Generally, methods listed under a) to e) are implemented in the standard workflow of the climate data section of NMHs. However, these quality control procedures are always a compromise between strength of regulations and acceptance of outliers. Moreover, everyday practice in data handling and database management still produces erroneous data after passing the quality control procedures mentioned above. Therefore, additional efforts are necessary to improve the data quality, especially for longer-term climate data series. Such efforts should not only include a quality check according to a) to e) but also a validation against independent data as, for example:

– measured discharge against discharge at catchment level from hydrological model forced by meteorological observations;

– measured snow water equivalent or snow height against modeled data from a snow cover model forced by precipitation and air temperature;

– measured glacier mass balance against modeled data from glacier mass balance model forced by meteorological data.

Such independent validation is a powerful tool to show, for example, consistency between air temperature, precipitation, and snow height or snow water equivalent. Discrepancies can be, for example, well associated either to problems in air temperature or precipitation measurements. Similar validation approaches were also applied to climate reanalysis data.

1.2.3. *Methods for climate data homogenization*

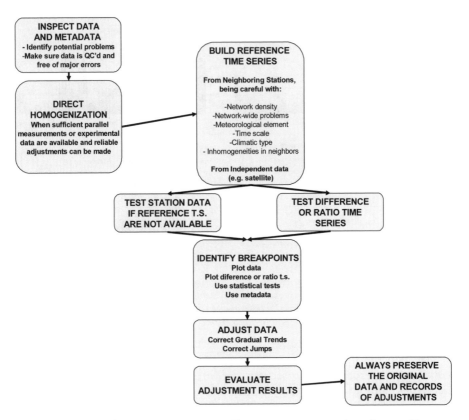

Figure 1.9. *Schematic representation of homogenization procedures for monthly to annual climate records (from [AGU 03])*

Whereas DQC is a well-established part of the standard workflow for the database of NHMs, data homogenization still has a strong link to climate research and not to standard procedures. This originates, in part, from the fact that powerful homogenization methods were only developed during the last approximately 15 years. Today there exists various types of homogenization tools for climate data involving different homogenization philosophies and with different strengths and weaknesses regarding climate element, geographical region, or temporal resolution of the dataset ([AGU 03]). It can be concluded from literature (e.g. [AUE 07],

[AGU 03]) that various homogenization tools work well for monthly time series, but homogenization of daily series or series with even higher temporal resolution is still not solved properly.

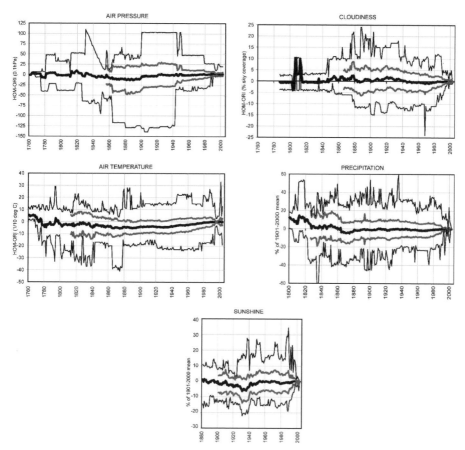

Figure 1.10. *Inhomogenities in monthly data series of air pressure, cloudiness, air temperature, precipitation and sunshine for the GAR (from [AUE 07]). Black bold lines show the mean of inhomogenity series, which implies that even averaging over many series cannot level out inhomogenities*

As previously outlined climate time series homogenization includes two important steps: first, the detection of breaks in the series (which need the creation of a reference time series), and second, data adjustment. Homogenization should be based on available metadata, which means that breakpoints should, as much as possible, be reflected by metadata information. However, many inhomogenities of climate time series are not captured by metadata (only perfect metadata would include all breaks). If available, parallel measurements from sensors, station locations, etc, should be used for adjustments. Today all homogenization tools for breakpoint detection use relative homogenity tests, which mean that they detect inhomogenities from statistical evaluation of the differences or ratios to a reference series computed from neighboring stations in similar climate conditions considering statistical significance of breakpoint. Absolute homogenity tests fail to differentiate between climatic and non-climatic inhomogenities.

Figure 1.9 summarizes the procedure of climate time series homogenization used by most of the homogenization tools used in climatology. Also the procedure varies with the different homogenization tools; however, it is laborious work, independent of the method used. Reference series are not only used for detection of breakpoints but also for adjustment of breaks. A useful summary of available homogenization methods (for both the breakpoint detection and the adjustment procedure) is given by [AGU 03].

It is important for homogenization work to consider the underlying statistical distribution of data to be analyzed. Non-normally distributed datasets need other approaches compared with datasets with normally distributed data.

Figure 1.10 shows results from homogenization of a multiple climate dataset for the GAR. It can be concluded from this figure that climate networks contains systematic changes of the network inherent to all stations and that averaging over a larger sample of climate series can not remove inhomogenities.

1.3. Metadata: documenting quality and usability

1.3.1. *Short characteristic of metadata from a climate data perspective*

Information on data, known as metadata, enables the operator of an observing system to take the most appropriate preventive, corrective, and adaptive actions to maintain or enhance data quality. Metadata, therefore, include detailed information on the observing system itself and, in particular, on all changes that occur during the time of its operation. Within the WMO, the role of ISO 19115 is stressed as the metadata standard for describing meteorological and climatological data. This standard has been developed for the geographical community, but it is also useful for meteorology as both disciplines deal with spatial data. Whereas the documentation of metadata related to the quality of the data is well established in meteorology and climatology, the description of the context of the data (e.g. access rules) is much less developed. Detailed and standardized metadata are increasingly important in the case of data access via the internet.

Elements of a metadata database (according to [WMO 08]):

A metadata database contains initial set-up information together with updates whenever changes occur. Major elements include the following:

(a) Network information:

(i) the operating authority, and the type and purpose of the network;

(b) Station information:

(i) administrative information;

(ii) location: geographical coordinates, elevation(s);

(iii) descriptions of remote and immediate surroundings and obstacles. (It is necessary to include maps and plans on appropriate scales);

(iv) instrument layout;

(v) facilities: data transmission, power supply, cabling;

(vi) climatological description.

(c) Individual instrument information:

(i) type: manufacturer, model, serial number, operating principles;

(ii) performance characteristics;

(iii) calibration data and time;

(iv) sitting and exposure: location, shielding, height above ground;

(v) measuring or observing program;

(vi) times of observations;

(vii) observer;

(viii) data acquisition: sampling, averaging;

(ix) data-processing methods and algorithms;

(x) preventive and corrective maintenance;

(xi) data quality (in the form of a flag or uncertainty).

A simple template for the description of station exposure is shown in Figure 1.11. As ideal exposure of sensors, according to the WMO standards, is seldom available, the documentation of exposure is of high importance for the usage of data retrieved from a particular station, otherwise the reliability of the observations cannot be determined.

Metadata on meteorological data are very important because they are highly dependent on observational practices, such as type of sensor, its exposure, observational procedures, observation times, etc. The full potential of data can only be explored when sufficient metadata are available; As stated by the WMO [AGU 02]: "The details and history of local conditions, instruments, operating procedures, data processing algorithms and other factors pertinent to interpreting data (i.e. metadata) should be documented and treated with the same care as the data themselves".

An important part of metadata is information on the local environment. Geographical coordinates and elevation of a station does not provide enough information in this context. Even if coordinates are documented with sufficient accuracy, the local environment can not be reconstructed. In particular, influences on station measurements act at different spatial scales. Therefore, documentation of the local environment should cover [AGU 03]:

– updated mapping in some form of the mesoscale region at ca. 1:100,000;

– toposcale map (ca. 1:5,000), updated each year, as specified by the WMO Technical Commission for Instruments and Methods of Observations [WMO 96];

– radiation horizon mapping, updated each year;

– photos taken from all points of the compass and sufficient area of the enclosure and of instrument positions outside the enclosure, updated upon significant changes;

– a microscale map of the instrument enclosure, updated when individual instruments are relocated or when other significant changes occur.

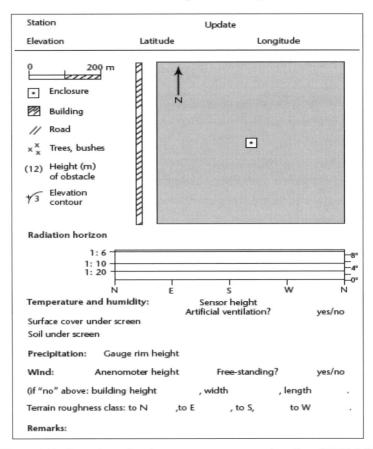

Figure 1.11. *General template for station exposure metadata (from [WMO 96])*

A detailed picture of metadata information can be derived from Figure 1.12 for the example of the climate station at Sonnblick in the Austrian Alps. It is clear from this figure that many changes occur within short intervals, although this station is outstanding with respect to station relocations showing no changes over the entire observation period.

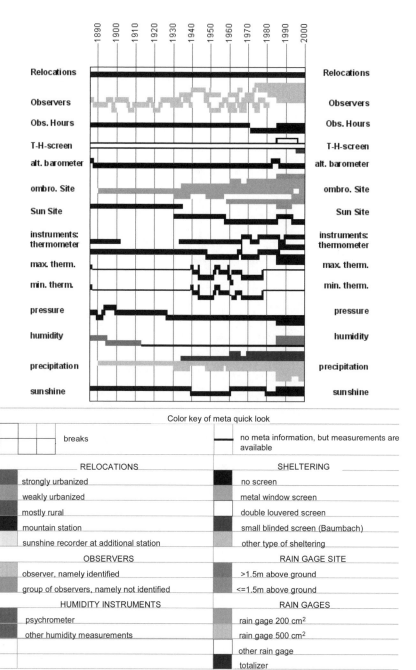

Figure 1.12. *Meta quick-look for the climate station Sonnblick (Austrian Alps) (from [AUE 01]), (see color section)*

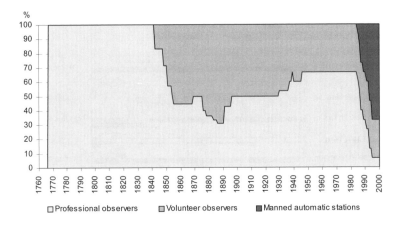

Figure 1.13. *Time series of the urban influence on the climate
network of Austria (from [AUE 01])*

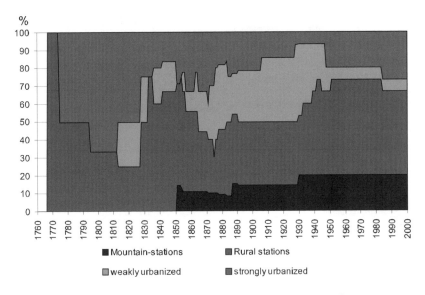

Figure 1.14. *Observers of the Austrian climate network
1767-2000 (from [AUE 01]), (see color section)*

Figures 1.12 shows the documented station environment within the Austrian
climate network since the beginning of systematic weather observations. It shows
the decreasing trend in the number of urbanized stations in the network compared
with rural stations. The systematic change from manned to automatic stations can be
seen in Figure 1.13. Today volunteer and professional observers are available only
at a very limited number of climate stations.

1.3.2. *Catalog services*

Metadata are an important basis of catalog services. The OGC Catalog Service defines common interfaces to discover, browse, and query metadata about data, services, and other potential resources. Providers of resources use catalogs to register metadata that conform to the provider's choice of an information model (e.g. descriptions of spatial references, thematic information). Catalog services are either web-based or client applications that can be searched for geospatial data and services in very efficient ways. Catalog services need standardized metadata according to ISO. The most important catalog service in Europe is INSPIRE the European geospatial data infrastructure. Catalog services will increase in importance in meteorology and climatology, but will need considerable effort in metadata servicing and handling.

1.4. Future perspectives

The integration of meteorology and climatology into GIS is an ongoing, extremely dynamic process, for which future perspectives are hard to predict. However, it can be assumed that meteorology and climatology will strongly benefit from integration into the GIS community because of the technical progress associated with GIS. In particular, the exchange and transfer of data with users and across disciplines will be significantly improved, and thus, it will be much easier for weather and climate models to integrate relevant (and updated) datasets from other disciplines (e.g. land use, soil, vegetation). The strong innovation and technological progress in GIS and emerging tools and applications will even stimulate the already well-established tools for weather forecasting in the future. The pressure towards web-based services will also force meteorology and climatology to more standardized metadata according to ISO standards. Finally, there is currently also great interest from GIS industry to incorporate meteorology and climatology and especially considering the special demands (in the OGC standards) from these disciplines.

Spatial data infrastructures like INSPIRE will additionally push forward the integration of meteorology and climatology into GIS. Geospatial data interoperability will soon include climate and weather data. OGC standards like WMS, WCS, WFS and new standards to be developed will soon fully be adopted by meteorology and climatology, and thus, enable the incorporation of meteorological and climatological data into different services. However, an important drawback still exists because of the different data policy of the various data providers (NHMs in particular), which could be a significant barrier to further development. Aside from this impedance, Shipley [SHI 05] stated that these parallel universes are already joined, or will soon unite in the future.

1.5. Bibliography

[AGU 03] AGUILAR E., AUER I., BRUNET M., PETERSON T.C., WIERINGA J., *Guidelines on Climate Metadata and Homogenisation.* WMO/TD No. 1186, World Meteorological Organization, 2003.

[AND 08] ANDERSSON E., THEPAUT J.N., "ECMWF's 4D-Var data assimilation system – the genesis and ten years in operations", *ECMWF Newletter*, no. 115, pp. 8-12, Spring 2008.

[AUE 01] AUER I., BÖHM R., SCHÖNER W., "Austrian long-term climate 1767-2000 (multiple instrumental climate time series from Central Europe), *Österreichische Beiträge zu Meteorologie und Geophysik*, vol. 25, p. 147.

[AUE 05] AUER I., BÖHM R., JURKOVIC A. *et al.*, "HISTALP – Historical instrumental climatological surface time series of the greater alpine region 1760-2003", *International Journal of Climatology*, vol. 27, pp. 17-46, 2005.

[AUE 07] AUER I., JURKOVIC A. *et al.* "A new instrumental precipitation dataset in the greater alpine region for the period 1800-2000", *International Journal of Climatology*, vol. 25, no. 2, pp. 139-166, 2007.

[BOH 98] BÖHM, R., "Urban bias in temperature series – a case study for the city of Vienna", *Climatic Change* vol. 38, pp. 113-128, 1998.

[CHR 02] CHRISTAKOS G., BOGAERT P., SERRE M., *Temporal GIS. Advanced Functions for Field-based Applications*, Springer Verlag, 2002.

[ESR 09] ESRI ArcGIS 9.2, 2009, online at: http://webhelp.esri.com/arcgisdesktop/9.2/.

[HIE 09] HIEBL J. *et al.*, "A high-resolution 1961-1990 monthly temperature climatology for the greater alpine region", *Meteorologische Zeitschrift*, vol. 18, no. 5, pp. 507-530, 2009.

[ISO 93] ISO, *International vocabulary of basic and general terms in metrology (VIM)*, International Organization for Standardization, Geneva Switzerland, 1993.

[KLE 08] KLEIN E., PERSSON C., "Showcase EUROGRID – towards a European resource for gridded climate data, products and services", EMS8/ECAC7 Abstracts, Vol. 5, EMS2008-A-00341, 2008

[LOW 09] LOWE D., WOOLF A., LAWRENCE B., PASCOE S. "Integrating the climate science modelling language with geospatial software a services", *Int J Digital Earth*, vol. 2, suppl. 1, pp. 29-39, 2009.

[NAT 04] NATIVI S., *et al.*, "Differences among the data models used by the geographic information systems and atmospheric science communities", *Proceedings of the 84th AMS Annual Meeting*, Seattle, USA, 11-15 January 2004.

[SHI 05] SHIPLEY S.T., "GIS applications in meteorology, or adventures in a parallel universe", *In Box Insights and Innovations*, American Meteorological Society, pp. 171-173, 2005. Online at: http://www.gmu.edu/departments/geog/People/Shipley/wxproject/papers/BAMS/shipley%20BAMS%20Feb%202005.pdf

[SZE 04] SZENTIMREY T., BIHARI Z., "Meteorological interpolation based on surface homogenized data bases (MISH)", in Szalai S., Szentimrey T., Lakatos M. (eds.) *Proceedings of the Conference on Spatial Interpolation in Climatology and Meteorology*, pp. 17-27, COST Action 719 – The Use of Geographic Information Systems in Climatology and Meteorology, Belgium.

[VAN 08] VAN DER VEGTE J., "Atmospheric data access for the geospatial user community", *Workshop on the use of GIS/OGC Standards in Meteorology*, ECMWF, 24-26 November 2008.

[WEL 05] VAN DER WAL F.J.M., "Spatial data infrastructure for meteorological and climate data," Meteorol. Appl., vol. 12, pp. 7-8, 2005, doi:10.1017/S13504827040011471.

[WMO 96] WMO (WORLD METEOROLOGICAL ORGANIZATION), *Guide to Meteorological instruments and Methods of Observation* (6th Edition), World Meteorological Organization, Geneva, Switzerland, 1996.

[WMO 07] WMO (WORLD METEOROLOGICAL ORGANIZATION), *Guide to Climatological Practices* (draft 3rd Edition), World Meteorological Organization, Geneva, Switzerland, 3 May 2007.

[WMO 08] WMO (WORLD METEOROLOGICAL ORGANIZATION), "Guide to meteorological instruments and methods of observation", *Weather-Climate-Water*, WMO no. 8, Geneva, Switzerland, 2008.

[WOO 05] WOOLF A. CRAMER R., GUTIERREZ M., VAN DAM K.K., KONDAPALLI S., LATHAM S., LAWRENCE B., LOWRY R., NEILL K., "Standard based data interoperability in the climate sciences", *Meteorol. Appl.,* vol. 12, no. 1, pp. 9-22, 2005.

[WOO 06] WOOLF A. *et al.*, "Data integration with the climate science modelling language", *Geophysical Research Abstracts*, vol. 7, 08775. SRef-ID: 1607-7962/gra/EGUOS-A-08773, 2006.

Chapter 2

Spatial Analysis, Cartography and Climate

2.1. Introduction

2.1.1. *Climatological information*

Climatological information is recorded in geographic space thanks to the use of weather stations equipped with different types of sensors, such as thermometers, rain gauges, anemometers, wind vanes, hyrgrometers, etc. There are two main reasons that explain why a global weather observation network has been developed:

– first, to provide the necessary information for weather forecasts. Weather forecasting is carried out by researchers and engineers who use data collected from all over the world to provide complex digital models. These complex digital models belong to the world of physics and include branches of physics such as thermodynamics, and fluid dynamics in particular. Due to these models, it has now become possible to use sets of differential equations to give an accurate description of atmospheric conditions and atmospheric activity. This aspect belongs to meteorology and is not dealt with in this chapter;

– second, to clarify which factors are required for a greater understanding of the climate, and how the climate functions in respect to both time and space.

Data that are collected from stations are used for climatological studies. Simple calculations can be used to recreate a continuous series of data that provides information on temperature, precipitation, wind speed, and direction, etc., on a daily, monthly or yearly basis. For a number of years now this climatological information has been closely examined because they provide us with information on the actual climate change.

Different variations of climate factors are not recognized from space on a continual basis.

Chapter written by Daniel JOLY.

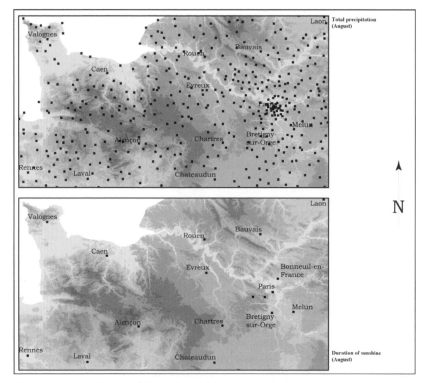

Figure 2.1. *Distribution of climatological stations in which total precipitation amount and duration of sunshine are recorded (data source: Météo-France)*

Trying to reproduce a continuous series in space has proved to be a more complicated task than was first thought. Climatological data are recorded on the Earth's surface through networks of meteorological stations. However, the density of these stations depends on the country and regions in which they are located, as well as on the variables that are to be studied. For example, in mainland France the average distance between climatological stations that record rainfall is 13 km, the average distance between those that record temperature is 20 km, and for those that record duration of sunshine this distance is 70 km. Figure 2.1 shows that spatial sampling tends to be irregular (the region of Paris is over-represented), and varies depending on the climate factor in question. For example there are 3,165 climatological stations that record rainfall, 1,530 climatological stations that record temperature, and 115 climatological stations that record duration of sunshine. This irregular distribution of climatological stations is partly a result of the complexity of the different environments in which the climatological stations are located. In addition to this, the irregular distribution of these climatological stations can also be linked to the importance that is given to a particular climatic factor that needs to be investigated, to their spatial variability, or also to the cost of the stations required to record the data.

Representing such data on a map poses problems. Theoretically, the only valid method of representing this punctual data on a map involves superimposing the results of the measurements onto the precise area from where the measurement was taken (Figures 2.1 and 2.5). However, this method of cartography does not provide any further information about the distribution and quantity of rainfall in France. This type of cartographic representation does not show any information on what happens in the areas between the measurement stations.

2.1.2. *Geographic information systems (GIS) and spatial analysis*

Spatial analysis, which is carried out by GIS, makes it possible to overcome the problems described above. The principle here is that the information that is available (the information type in question is the first point that needs to be examined) contains both random variables and regularities. The term random is used here to describe everything resulting from the inaccuracy of a measurement, from errors that occur during the recording of the information and from the transmission of this information. This random part of the information cannot be modeled. Conversely, the term regularity can be translated by gradients and these can be modeled, if the tools that are required to model them are available. The tools required to model the gradients are the second point that needs to be examined. Variation in precipitation linked to altitude is a trend that exists almost everywhere. Other trends also exist and they need to be identified, and this is what the technique known as interpolation aims to do. Interpolation makes it possible to recreate continuous spatial fields from punctual information.

Spatial climatology requires that a piece of information be as complete as it possibly can, and that the information be gathered and organized in such a way that it can be used by specific tools for analysis. The results of any analysis could, therefore, be used in agronomic models or represented on a map. The interpolation of the climatological data belong to the GIS field (raster GIS to be more precise). A raster GIS is most suitable system that can be used because it can deal with the problems that are associated with spatial analysis. The aim of this chapter is not to provide an analysis of the interpolation technique, as there is a lot of material already available that deals with this subject. The aim of this chapter is to provide an understanding of the problems that arise when geographic information is used during the interpolation of climate data. Other issues that this chapter will address include: what spatial resolution is the most suitable for providing the best results? How does one choose the best independent variable or variables? What method of interpolation should be used?

– For example, is it best to have a piece of information rasterized at a rough resolution (250 m, 500 m, or even 1 km), which is available on a large scale at a low cost, rather than having a piece of information with a smaller resolution (50 m), which is more expensive and difficult to manage if the area to be investigated is large? What impact would this spatial information have on the results of climatic interpolations? It is quite tempting to believe that the quality of the results is proportional to the accuracy of the information used to produce them. What really happens? Does this so-called loss in quality (associated with large-scale resolutions) occur to the same extent for the interpolation of both average and daily data? This raises the issue of spatial-temporal scales, which are an integral part of interpolation.

– There is also the issue of the selection of independent variables, which must be quantitative this is the case for digital elevation models (DEM). However, as far as land cover is concerned (when using Colrine land cover (CLC), how is it possible to do with the technique of interpolation when the only variables that are available are qualitative?

– The choice of interpolation method to be used depends on numerous factors, such as the type of variables to be interpolated, the information that is available, etc. A brief overview of all the different interpolation methods will be given in this chapter, with the help of specific examples. Unfortunately, however, our research will be unable to provide a miraculous solution to those climatologists and geographers who want to use a specific interpolation method, but it will help them choose a particular method so that they can continue to carry on with their own research or work.

All of the issues raised here are based on new processes that were developed from data recorded and published by Météo-France. The data collected related to rainfall, temperature, and duration of sunshine in mainland France.

2.2. Geographic information necessary for interpolation

It is not possible to represent information on a map if nothing is known about the location of the measurement stations, or if there is no information available about the data that is to be represented. The third important element that also needs to be taken into consideration refers to information that exists outside the world of climatology and which is used as a support to spatial analysis.

2.2.1. *The location of the measuring sites*

It is possible to find out the exact location of each weather station on the Earth's surface due to the geographic co-ordinates that make-up the geometric aspect of a GIS. Until the middle of the 1990s it was only possible to work out the location of such climatological stations by using topographic maps. Measurements taken from the stations tended to be full of errors. A well-trained geographer is able to locate a meteorological station on a topographic map to the nearest millimeter, which translates as being 50 m on a map with a scale of 1:50,000 or 25 m on a map with a scale of 1:25,000.

Such methods for working out the locations of weather systems have been replaced by GPS systems. Even if the potential margin for error has decreased, it should be highlighted that the inaccuracy of a measurement can be approximately

10 m with a low cost GPS (bi-frequency professional GPS are exact to the nearest millimeter). This inaccuracy is of no importance if the interpolation has been carried out with a rough resolution raster GIS. However, such an inaccuracy can cause problems if the size of the inaccuracy is actually greater than the resolution size. Figure 2.2 shows an example of two rasters with a spatial resolution of 100 m and 10 m, respectively and a positioning inaccuracy of 15 m, which is the average error for a low-quality GPS.

Regarding the 100 m raster, the inaccuracy of 15 m in positioning does not cause any location error when the weather station is located at or near the center of the pixel: the area of uncertainty in which the points (provided by the GPS) can be found are all located quite far from the pixel's boundary limits. Irrespective of the GPS error, the location of the weather station will always be towards the center of the pixel (example A in Figure 2.2). However, whenever the true location of the weather station gets closer to the pixel's boundary limits, the total area for inaccuracy increases and also becomes closer to the pixel's boundary limits (B1). In this example, the location of the weather station in the correct pixel is not necessarily affected. This situation can quickly change whenever the station is located at least 15 m from the pixel's boundary limits. This occurs due to the fact that there is a non-zero probability that the GPS will provide a location that is outside the correct pixel (B2). The minimum probability of 0.5 is possible for a weather station that is located at the pixel's boundary limits (B3). Therefore, with an inaccuracy of 15 m and a resolution of 100 m, the probability that a weather station will be placed in the correct pixel is 0.49 (this is the area covered by the square with dashed lines in Figure 2.2b). The probability that a weather station will be placed outside the pixel is 0.13.

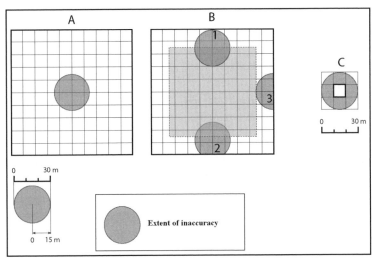

Figure 2.2. *Problems associated with location inaccuracies with a 100 m (A and B) and 10 m (C) raster*

In the other example where a 10 m raster is used (example C in Figure 2.2) the situation is worse because the size of the inaccuracy is bigger than the size of the pixel. The exact point of location (depending on the moment when the measurement was taken) may be in one of the nine pixels that surround the theoretical center point from where the measurement was taken. In this example, a clear error has been produced.

2.2.2. *Endogenous information*

Endogenous information relates to all climatological data that can be mapped. Many organizations, in France and in other countries, measure, archive and transmit climatological information. Météo-France is the only organization in France that is entirely dedicated to meteorology and climatology. Météo-France makes several of its databases available to the general public in the form of its "CLIMATHEQUE", which is the only source of homogenous information that has been verified and validated and which is available throughout the whole of France.

Within the CLIMATHEQUE it is possible to access basic data, such as precise hourly, daily, 10-day or monthly data, etc. The different variables that are available include temperature (maximum and minimum daily temperatures, average temperature or frequency of high (>30°C) or low (<–5°C) temperature) precipitation (frequency, amount), sunshine, frequency of fog, windspeed, etc. Other types of data, which are the result of specific processes, are also available and they include climate tables, calculations of degree-days, which are to be issued to the École Spéciale des Travaux Publics (ETP), a prestigious university in Paris. Only the basic data will be used in the following examples.

2.2.3. *Exogenous information*

Ever since the end of the 1980s exogenous information as information layers, covers all space and is capable of explaining the spatial modifications being experienced by different climate variables. Transferring information using spatial analysis tools makes it possible to recreate continuous spatial fields from precise endogenous data.

These layers of exogenous information are stored within the GIS and make extensive use of digital elevation models (DEMs) as well as remote sensing which are the two main initial data sources.

2.2.3.1. *DEMs*

The DEM is an altitude matrix that is provided in France by the French National Geographic Institute (IGN). The database in question provides a DEM with different resolutions which range from 50 to 500 m, even 1 km. In this chapter, we will use different resolutions due to the fact that the climate models in the following examples using different scales. By calculation applied to a DEM, it is possible to derive many types of information layers. In order for this to take place, each of the pixels of the DEM is successively considered as being the center of an n×n pixels window. The altitude values of this window are adjusted by regression to obtain a plan from which it is possible to calculate different topographic variables such as:

– altitude (value of the window's central pixel);

– gradient of the slope, which corresponds to the incline of the adjustment plan that is to be carried out on the window;

– orientation of the slope;

– topographic roughness, which corresponds to the standard deviation of the residuals within the adjustment in altitude; this topographic roughness highlights the importance of the irregularities in relief that exist (in flat regions or on a straight mountain face there may be no irregularities in relief);

– the theoretical model for global solar radiation is calculated for different key moments of the year (summer or winter solstice, spring and fall equinox), and also takes certain hidden topographic factors into consideration (e.g. the effect that shadows have on the model is particularly significant, especially in winter).

In this way, other information layers can be provided by the DEM.

2.2.3.2. Land cover

Land cover, from satellite or aerial images, provides information on how the landscape is made up of different classes of objects (vegetation, developed sites, networks, crops, etc.). Land cover is the second most useful, if not essential, source of information used to model the climate in its different spatial variations. With the use of relevant calculations, it is possible to derive other data from information relating to land cover:

– with the help of TM imagery it is possible to calculate the Normalized Difference Vegetation Index (NDVI). This is the index that estimates biomass potential. The NDVI is calculated from channels 3 and 4 of the Landsat TM images by using the following formula: (TM3-TM4) / (TM3+TM4);

– from a simple classification it is possible to calculate the distance that exists between each pixel of different objects such as the sea, the forest or the towns.

The land cover of a particular country can also be provided by other sources of information. CLC (supplied by the French Institute for the Environment) observes land cover on a scale of 1:100,000. The main aim of CLC is to respond to both French and European needs. The minimum surface area that can be considered (description threshold) is 25 hectares; the minimum width of the network elements that are to be observed (hydrography in particular) is 100 m. This information, is suited to a raster GIS with a resolution of 250 m and could be if it is a question of modeling small-scale climatic spatial variations, for example the variations that occur in France.

All of these information layers are stored in a GIS and will be used in analysis as explanatory variables. All of these information layers are not necessarily required for the interpolation process to take place. The variables that are available will influence any decision on the interpolation method that is to be used.

2.3. The main interpolation methods

The climatological data are located in what is known as 2D space (a third dimension can be added if altitude is taken into consideration; however, in a GIS altitude is considered more as an attribute than as a dimension). The theory associated with spatial analysis involves providing an explanation about the localization of the geographic objects that exist in space [ARN 00]. There are two major stances that can be adopted as far as interpolation is concerned [CAR 03]:

– the first stance bases interpolation on spatial autocorrelation; the techniques of interpolation that are used rely only on the location of the measurement sites;

– the second stance introduces geographic information in a different way, by making explicit reference to space but in a different context. What is important here is not the location of a measurement station in relation to its neighbors, but rather the environment that surrounds the measurement station. In this way, it is the geographic objects that are continually distributed in space and which can be distinguished thanks to the shape and form of localized characteristics (exogenous independent variables such as latitude, longitude, altitude, slope, etc). These geographic objects will then be used as a support for the interpolation process.

2.3.1. *Interpolation based on spatial autocorrelation*

The fundamental issue here relates to spatial correlation techniques, which are based on distance and are well suited to the interpolation of localized data. The fact that autocorrelation exists makes it possible to carry out analyses such as the techniques of interpolation. This means that a particular piece of data is similar to its neighboring data, and the similarity is inversely proportional to the distance that seperates them. This inverse proportion, which can take many forms, leads to the creation of gradients, which is another geographic concept. In all cases, autocorrelation characterizes measurement stations, or neighboring regions whose measurement values are similar [FLA 01; PHI 01]. If the measurements for a particular given space were distributed randomly in terms of the distance that separates the measurement sites, then it would not be possible to carry out spatial analysis based on the process of autocorrelation.

It is possible to use three main methods of interpolation when only its x and y co-ordinates (longitude and latitude) are known: inverse distance weighted (IDW), kriging, and cubic spline. Each interpolation method has its own advantages and disadvantages. The three methods rely on the idea that the value of any localized measurement site depends on the values of the surrounding measuring stations [GRA 02]. The main differences between these three interpolation methods are the way in which the different variations of a station's value is modeled, and the way in which weighting factors are allocated. Weighting factors are used to allocate values to pixels that originally have no value [MER 01]. A brief overview of these three methods will be given in the next part of the chapter and examples that refer to the spatial variation of sunshine in France for the month of July will be used.

2.3.1.1. *Inverse distance weighted interpolation*

Newton's theory of gravitation is applied to the technique of IDW interpolation. The principle states that the influence of a measurement that is taken from any site in space decreases inversely with distance [LAN 94; LAN 96]: this means that a site that is located near a weather station, from which a measurement has been taken, will have a value similar to the values measured in the station and the inverse is true for sites that are located further away from such climatological stations. The weighting factor is provided by the taking the inverse of the distance or possibly by taking the inverse of its square or log.

2.3.1.2. *Spline Interpolation*

Different spline methods exist [ECK 89; MIT 99]. A spline interpolates a continuous area from a local sample of sites by using a minimum curve, which means that this continuous area will include the sample sites. The most common spline method used is the one in which local values are adjusted by cubic polynomials. The model seen in Figure 2.3b was created with the help of the standardized function provided by the software ArcGIS. There are 12 sample sites and the weight allocated to each of the sites is 0.1.

2.3.1.3. *Kriging*

Kriging is a stochastic interpolation method that can also be referred to as an optimal method or as a method of objective analysis, depending on the field it is used in. Just like the previous interpolation methods mentioned, the kriging approach takes full advantage of spatial autocorrelation. The aim of the kriging method is to identify the autocorrelation area which is limited by an optimal distance. This distance, which corresponds to the vertex of a variogram, acts as the boundary between the following two samples:

– the first sample is located within the boundary and groups together all of the sites that possess similar spatial organization characteristics;

– the other sample, located outside this range, is made up of stations that are characterized by random variations.

Kriging is quite a complicated process and demands a good understanding of spatial statistics [BUR 86; OLI 90]. Kriging is based on the theory of regional variables. This theory states that the spatial representation of any variable, which can be represented by quantitative values, is homogenous for any given surface. This means that the same spatial distribution model must be observed in the whole studied area. Local spatial variation is modeled by the semi-variogram, which describes how data functions in relation to space.

The estimation model, which is calculated from the variogram, is introduced as an equation which determines and fixes the weight that needs to be applied to the data during the interpolation process. The kriging method has frequently been used in many different applications within the field of climatology [COU 99; LAB 95]. In order to create Figure 2.3c, the standard function from the ArcGIS software program was used. Within this particular function the normal kriging method comes equipped with a spherical semi-variogram.

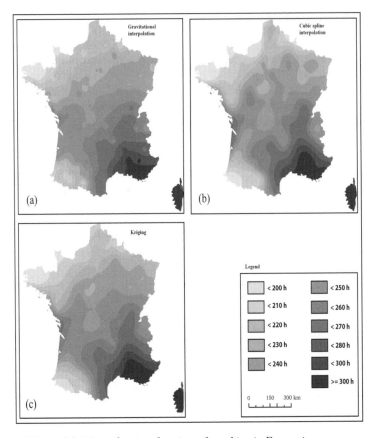

Figure 2.3. *Maps showing duration of sunshine in France in summer. These maps were produced by the three different interpolation methods: a) gravitational, b) cubic spline, and c) kriging*

2.3.1.4. *Maps generated by the three different interpolation methods*

Figure 2.3 shows maps that were generated by the three different interpolation methods mentioned (IDW, cubic spline, and kriging). These maps show the duration of sunshine in France in summer. Each map has produced very similar results with those for the Mediterranean area to the south contrasting with the results for coastline between Brittany and the region of Nord-Pas-de-Calais. Differences appear, particularly in areas such as the Western Pyrenees, where the cubic spline and kriging methods have produced results that show a lower amount of sunshine (less than 200 hours) than those produced by the IDW method.

The majority of authors who worked on this research [LAS 94] believe that whenever a regular sampling grid is used, the cubic spline method tends to produce results that are equivalent to those produced by the kriging method [DUB 84]. The cubic spline method is quicker and much easier to use.

The IDW method requires a substantial amount of processing time if there are several thousand points that been allocated values. The results produced by IDW interpolations [COL 00] were the most unreliable.

Interpolation methods, which are based on functions where distance is the only estimator variable, are normally considered as being low-quality interpolation methods in the world of climatology. It is true, that two neighboring climatological stations will generally have the same or a similar temperature. However, when a more detailed analysis is carried out things are not as simple. Temperature, as well as all of the other climate factors (including sunshine), vary in space and this depends on factors that are associated with the topography of the Earth's surface or factors that are associated with land cover.

Such large error estimations can be corrected by the other important aspect of the interpolation process, which involves carrying out analyses by using continuous spatial information.

2.3.2. *Interpolations based on continuous spatial information*

This second type of interpolation is based on a deductive approach, in which the factors that produce the climate of a particular area have to be identified. This type of approach is acknowledged and appreciated by all those people who believe that it conforms to a certain determinist view of nature, an approach that is used to predict climate variables. There are many examples based on this approach that have been published in different papers on the topic.

Certain physical factors obviously need to be considered in this approach and they include altitude (in accordance with proportionality) which links air pressure with temperature. Other physical factors include gradient, which controls the flow of air along the mountain slopes, as well as aspect of the slopes, which exposes the mountain face to sun rays or to air flow, etc. Other physical factors provide information to explain the spatial variation of temperature or precipitation: distance from a particular geographic object, the fractal dimension of features such as wooded areas, etc. This is why it is common to use a set of factors that have an influence on the spatial variation of climate factors.

These spatialized factors form part of the interpolation functions as external drift (in the case of kriging) or as independent variables (in the case of statistic interpolations). The factors are archived as information layers in the GIS. These two types of interpolation not only rely on the use of the independent variables, but also on the location of the climatological stations, which is essential.

2.3.2.1. *Kriging with external drift*

Kriging is a very reliable technique from a statistical viewpoint, because it is based on the analysis of covariances. One of the main criticisms that can be applied to the method of kriging is that the spatial variations of the phenomena that are analyzed are not only a function of spatial autocorrelation, but a function of other processes. In many cases, variables, which are weakly correlated with a phenomenon that is being analyzed, will have a strong influence on the spatial variation of the phenomenon in question. For example, slope and the topography of the Earth's surface have a considerable effect on temperature. The kriging method retains this information with the help of external drift.

In comparison to normal kriging, kriging with external drift improves the accuracy of estimations that have been made, especially in zones where there are not enough samples [HUD 94; TAB 05].

2.3.2.2. *Statistical spatial analysis*

Theoretically, correlations that are solved by the process of least square regressions are not appropriate when there is autocorrelation, in other words when there is spatial dependence between the connected variables. It is still not possible to avoid autocorrelation. Most attention is needed when such variables are to be integrated.

It is strongly recommended, if not necessary, to carry out a statistical test on variables [LHO 05].

This type of determinist approach can be explained in more detail with the help of Figure 2.4, which is an example of interpolation that uses spatial statistics. This approach relies on the use of diverse information, which is digitally archived and which is then combined together [FEY 95; JOL 03]. This approach is then used to locate information (from the information that is available) which would be suitable in helping to solve the problems associated with interpolation.

The principle of this approach involves calculating (by multivariate regression) the links that are established between the endogenous climatological variable that is to be interpolated, and the independent climatological variables that are stored in the GIS as information layers. The process of statistical modeling makes it possible to recreate continuous fields of endogenous variables.

The formalization of these links leads to the creation of a map operator. This map operator makes it possible to transfer knowledge that has been acquired from one particular piece of information to the entire area that is to be studied.

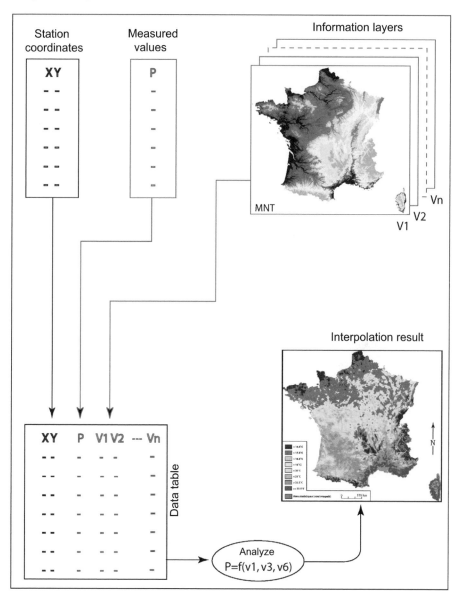

Figure 2.4. *Flow diagram showing how a determinist interpolation model works (see color section)*

2.4. Geographic information used in statistical interpolations: advantages and disadvantages

In addition to the problems dealt with in the previous section, other problems also arise and these problems limit or prevent the use of spatial statistics in the interpolation process. Some of them will be dealt with by studying the following four points:

– studying the problems linked to the density of climatological stations in certain areas;

– studying the information source and the quality of interpolation. In this case the following issue is raised: how is it possible to convert data that provides information on land cover into quantitative data that can then be used within the regression models;

– studying the problems that are linked to the use of certain estimators (for example, distance from a geographic object). These estimators are supposed to play a role in the spatial distribution of certain climate factors;

– spatial variables and scales.

2.4.1. *Density of climatological stations*

The density of climate data available influences the way in which the data itself is interpolated. If we look at Figure 2.1 again it is clear that the density of climatological stations, from where rainfall amounts and duration of sunshine was recorded, is completely different.

2.4.1.1. *Rainfall amounts*

Rainfall amounts are recorded by 3,165 rain gauges in France (see Figure 2.5). For recordings taken in the month of August over a 30 year period from 1971 to 2000, the average rainfall was 590 mm, with a series that ranges from 9.9 mm (in Corbara, Corsica) to 145.8 mm (in Lamoura, Jura). The large number of measurement points means that it is possible to analyze rainfall amounts with a high resolution because all of the spatial configurations are sampled.

Figure 2.6 compares the frequencies of 13 different altitude classes, and deals with two distinct populations:

– the first one is a corpus of 8,833,676 pixels with a resolution of 250 m and which covers the whole of France;

– the second one is a corpus of the 3,165 pixels with a rain station.

Only two of the different altitude classes produce slightly different results from one another. The altitude class, which includes altitudes of ≤50 m is over-represented (11% of all rain stations come from this class, whereas this altitude class makes up only 5.4% of the total altitude range in France). Conversely, the higher-level altitudes are under-represented (only 0.1% of all rain gauges come from the altitude levels that exceed 2,000 m, whereas this class of altitude levels make up 2% of the total altitude range in France). Isola (which is located in the extreme southeast of France) was the rain station that was located at the highest altitude (at 2,509 m). This means that there is currently no information available for altitude levels up to 2,509 m. Therefore, regarding climatology, altitude levels of more than 2,509 m can be considered as unknown space. The over-representation of lower altitude levels does not affect statistical analysis (this fact was proven by randomly removing the surplus rain station). On the contrary, it is not recommended to carry out a process of extrapolation in areas that have an altitude greater than 2,300 m due to the lack of data available for this altitude level.

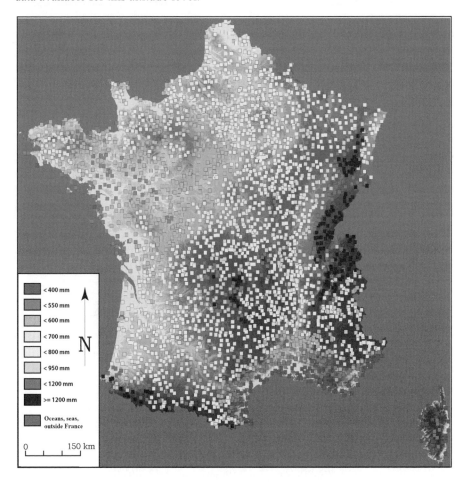

Figure 2.5. *Distribution of the 3,165 rain station that exist in France, and from where information on total rainfall is provided; August values (1971-2000 average). Source: Météo-France (see color section)*

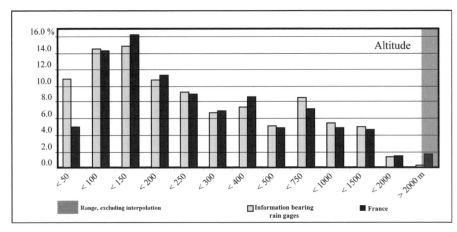

Figure 2.6. *Frequencies for the 13 different altitude classes in France
(8,833,676 pixels) and for the 3,165 rain stations in France*

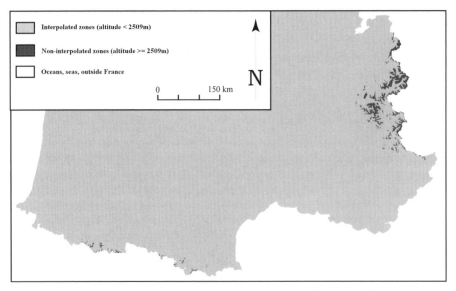

Figure 2.7. *A map showing the zones in which it is possible and in which it is
impossible to carry out the process of interpolation of rainfall amounts
according to the criterion of altitude*

With this in mind, the process of interpolation is not valid for many pixels that
are located in the Alps and the Pyrenees if the criterion of altitude is taken into
consideration (Figure 2.7). In the belief that interpolation is based on multiple
regressions, it is the set of predictive variables that need to be criticized. The aim of
this criticism is to find and mark out the areas in which the process of interpolation
is possible. The area in which the process of interpolation can take place normally

gets smaller as the number of predictive variables that exist within the interpolation models also gets smaller.

2.4.1.2. *Duration of sunshine*

Duration of sunshine, which is measured in hours, is recorded in only 115 evenly distributed climatological stations in France (Figure 2.8). This corpus of climatological stations only samples a small fraction of the sites that exist in France. Figure 2.9 shows that:

– the frequencies of the two groups are similar in relation to the classes of intermediary values (150-500 m),

– there is an over-representation of climatological stations that record duration of sunshine at the lower altitude levels,

– in comparison to the lower altitude levels, there is an under-representation of such climatological stations at the higher altitudes, for example, no weather station measuring duration of sunshine was found at an altitude above 1,020 m (Brenoux, Lozère).

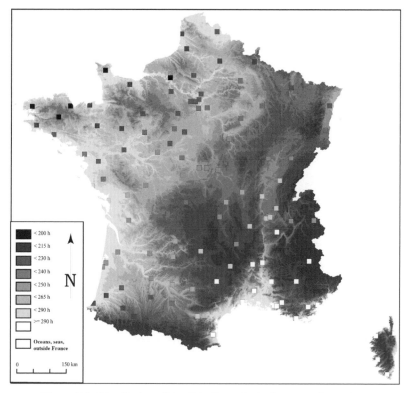

Figure 2.8. *Distribution of the 115 climatological stations from where duration of sunshine is recorded; August values (1970-2000 average). Data source: Météo-France*

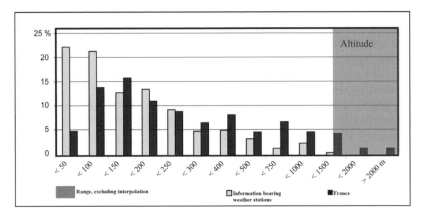

Figure 2.9. *Frequencies for the 13 different altitude classes in France (8,833,676 pixels) and for the 115 climatological stations which record duration of sunshine*

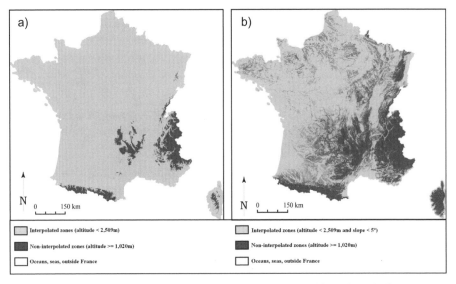

Figure 2.10. *Maps showing zones in which it is possible and in which it is impossible to carry out the interpolation of duration of sunshine, according to the a) criteria of altitude, and b) altitude and slope*

These restrictions are much more severe as far as slope is concerned: no sunshine station was found at a slope of more than 5°. Interpolation based on altitude limits the field of interpolation to altitude levels lower than 1,020 m (Figure 2.10a). If we consider interpolation based on altitude and slope, then the field of interpolation becomes narrow as it then becomes impossible to carry out the process of interpolation on slopes of more than 5° (Figure 2.10b). If altitude is considered as being a factor that does not have much effect on the duration of sunshine in summer ($r = 0.2$), then the slope of a mountain is considered as having a slight effect on the duration of sunshine in summer ($r = 0.46$). Under these conditions, the correlation that exists between the duration of sunshine (the dependent variable), and slope (the independent variable) does not make much sense in terms of space. This is due to the fact that the correlation between these variables is calculated from such a small-range data series (0 to 5°).

It is difficult, if not impossible, to estimate the duration of sunshine by using either linear or multiple regressions. In order to carry out such an estimation it is best to use the kriging approach as this approach provides the best results. This is due to the bad spatial distribution of the climatological stations that record the total duration of sunshine. These stations are generally synoptic and are found in uniform, homogenous sites that increase the level of autocorrelation. Let us notice that a kriging with an external drift does not provide any better results than the normal kriging method. Here the problem is more informational than related to the interpolation method used.

Several factors need to be determined before any process of interpolation can take place, for example the number of climatological stations where the climate variables are to be observed and recorded, the location of the climatological stations and the density of the climatological stations in a given area. The type of interpolation method which can be used depends partly on these factors.

2.4.2. *Information source and quality of interpolation*

In the field of spatial statistics the most difficult problem to overcome is collecting the independent variables. In the first part of the chapter we saw that there are two major types of geographic information that are readily available and they include DEMs and information resulting from remote sensing. The issue raised in this section refers to the effect of the information source on the quality of estimations. To help answer this question, data from two distinctly different sources of information will be compared, i.e. CLC and SPOT. Interpolation works well if the independent variables introduced into the regression models are of a quantitative nature. Variables, such as altitude, slope, and roughness of terrain, do not pose any problem. However, variables that provide information on land cover are more difficult to deal with. Measuring the distance that these variables are from an object is a useful and practical method to use when it comes to measuring the thermal and pluviometric influence that a particular object might have on its surrounding area. In some cases it is necessary to rely on the fractal dimension of certain objects, such as wooded areas or developed areas (areas with lots of sites on them), in order to estimate the influence that the spatial distributions of these formations will have on climate variables [FUR 94; JOL 03].

Another variable that is often used is the NDVI, which is easy to integrate within the different interpolation models. It should be pointed out that the NDVI is a simple numerical indicator that not only integrates data, such as type of vegetation, but also the state in which the vegetation can be found, such as lack of water, disease, infestation of insects, etc. The index reflects the importance that vegetation has in terms of influencing the variations of temperature and precipitation. In some cases, however, it is difficult or it may not be possible to obtain the NDVI, and this is especially true if the area to be interpolated is large. It is possible to overcome this problem if there is an image that shows land cover and if this image provides information on land cover. This is made possible by CLC, where the information provided is of a qualitative nature and cannot be easily integrated into the regression models. By using a CLC it is possible to obtain an index that provides information about the potential amount of vegetation that exists in each pixel of an image. The procedure for carrying out this task is relatively easy. In order to produce such a piece of quantitative information from CLC, and to measure how good this information is at estimating data, we will use the example of monthly temperature averages calculated for the period 1971-2000 by the 1,530 temperature stations in mainland France.

2.4.2.1. *The location of climatological stations and CLC*

The 1,530 climatological stations are projected onto a CLC image, which has been processed by a 250 m resolution raster GIS. The land cover type of each pixel with a climatological station is archived in a table that is made up of 1,530 rows (climatological stations) and 44 columns (one CLC post per land cover type). In the event where a climatological station and its corresponding CIC post cross paths, this case is given a value of one. All other 43 columns are given a value of zero. The sum of all of the rows is equal to one (each climatological station is characterized by only one land cover post). The sum of all of the columns provides information on the frequency of the land cover types for the 1,530 pixels where a temperature station is installed. This disjunctive table can be represented as a graph (see Figure 2.11).

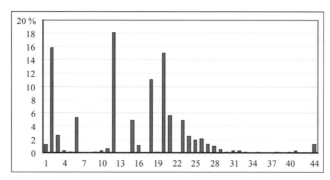

Figure 2.11. *Percentage of climatological stations located in each of the CLC's 44 land cover posts*

The climatological stations that record temperature are found on only four different types of CLC land cover with a frequency of up to 10%:

– type 2 (discontinuous urban areas): 16% of all climatological stations;

– type 12 (arable ground, outside irrigated areas): 18% of all climatological stations;

– type 18 (meadows): 11% of all climatological stations;

– type 20 (crops and partially cultivated land): 16% of all climatological stations.

Approximately 20% of all climatological stations that record temperature can be found on an additional four different types of CLC land cover:

– type 6 (airports): 5.2%;

– type 15 (vineyards): 4.8%;

– type 21 (mainly farm areas, interspersed with nature reserves): 5.5%;

– type 23 (leaved forests): 4.8%.

These eight different types of land cover represent, in total, more than 80% of locations of the climatological stations. Two of the land cover types correspond to built-up environments (types 2 and 6, which in total represent 21% of all stations). Open agricultural areas (types 12 and 18), as well as hedged farmland areas or more complex farmland areas that have a mixture of crops and different featured formations (types 20 and 21), represent 29% and just over 21% of all the stations.

It is also interesting to point out that almost 5% of all of the stations are found in forest areas. This is almost impossible because according to the rules drawn up for the localization of such meteorological stations it is not possible to place the stations in forest areas. These areas are largely dominated by woods, open spaces exist over a small surface area and the CLC does not recognize this.

Amongst the 36 other land cover types proposed by the CLC, 12 of them did not have any meteorological station. These 12 land cover types included: type 7 (extraction of minerals), type 14 (paddy fields), type 17 (olive groves), type 34 (glaciers), and type 43 (coastal lagoons).

2.4.2.2. How did temperatures react to CLC?

The table that has just been described is used as a physical support for a particular statistical process: correlating the temperature of the stations with the relevant type of land cover on which the station can be found. This process requires a large number of individuals. With the frequency of each land cover type not being high enough to cope with such a correlation analysis, amalgamations have been made. The following groups have been created with the aim of estimating the influence that each of the four main land cover categories has on temperature:

– built-up areas: includes all man-made areas (392 occurrences);

– complex zones: corresponds to all those areas that cannot be described as being cultivated or meadow (388 occurrences);

– open spaces: includes cultivated areas and meadows (402 occurrences);

– leaved forests (143 cases); coniferous forests have not been included because they might introduce some bias into the results of the correlations as they tend to be found in mountainous and thus, cold regions.

The overall result is quite conclusive: climatological stations that are located in an environment dominated by man-made objects (dense built-up areas, discontinuous urban materials, tarmac areas, etc) are associated with having higher temperatures than any of the other stations. There is a positive correlation coefficient, particularly for the months of April to July (inclusive), see Figure 2.12. Conversely, climatological stations that are located in areas that are recognized by CLC as being wooded areas (leafy) are associated with having lower temperatures than the other stations; there is a negative correlation coefficient for the months of June to September inclusive.

Open spaces induce negative coefficients all year around. In saying this, however, the coefficients are weaker in winter (similar to the coefficients produced by wooded areas) than in summer where the coefficient is closer to zero. This fact alone suggests that crops and meadows might have a slight influence on temperature. Farming zones have coefficients that are similar to the coefficients of an urban area in winter. As spring time progresses the coefficients of the farming zones become increasingly similar to those of open spaces.

Figure 2.12 provides information on the coefficients of the four different land cover types for each month of the year and mainly represents the influence that evapotranspiration has on temperature. During the winter months the different land cover types each have a high humidity level, which is due to high rainfall in winter and temperature decreases. The gap between the values of the coefficients for each land cover type is at its lowest during the winter, with all of the values being closer to zero.

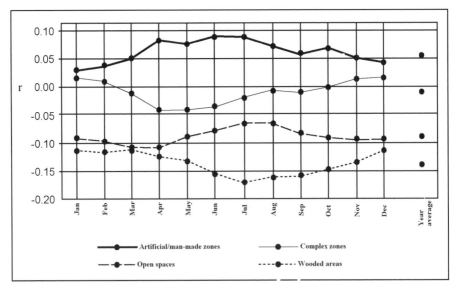

Figure 2.12. *Correlation coefficients between CLC land cover types for each month of the year*

As soon as spring arrives the gap increases again. Man-made surfaces, where asphalt surfaces dominate, become dry and heat up again quickly. It is at this moment that vegetation also starts to grow. Surfaces on which significant amounts of vegetation are present do not retain any heat due to the process of evapotranspiration. Temperature in these areas is lower in comparison to areas where there is much less vegetation. This trend continues into the middle of the summer for climatological stations that are located in woodland areas.

This same trend stops in April and then the reverse of this trend occurs until the month of August for CLC posts located in extensive open spaces. The growth of vegetation in open spaces is at its maximum in spring and as a result the processes of haymaking and harvesting make cultivated areas similar to mineral surfaces as far as temperature is concerned.

2.4.2.3. *Estimating the NDVI by transforming qualitative CLC values into quantitative values*

This model provides us with the tools required so that each of the CLC codes can be provided with quantitative values. Negative coefficients are associated with climatological stations found in relatively cool areas. These coefficients are given negative values because the biomass found in such cool areas is significant. A positive coefficient value is given to CLC land cover types that are part of the built-up area category where biomass is low. The reference value for each group of CLC land cover type is obtained by multiplying the annual coefficient by 100 (see Table 2.1).

	built-up areas	complex zones	open spaces	woodland spaces
Value	55	−10	−90	−135

Table 2.1. *Quantitative values allocated to the main land cover types as recognized by the CLC*

If a particular category corresponds to a similar CLC land cover type, only one quantitative value is allocated to this particular land cover type. For example, the group woodland spaces, which includes leaved forests and coniferous forests, has a value of −135, and there is no distinction made between the two types of forests. This is not the case, however, for areas composed of artificial surfaces (built-up areas, see Table 2.1) which are made up of approximately 10 different man-made types. The reference value of 55, which is the reference value for the entire group, will vary depending on the manmade material in question and in proportion to the actual amount of mineral surface that makes up the type. This is why a value of 110 was given to continuous urban material, in other words this part of the group has twice as much mineral surface in it when compared with the overall average for the entire group. Urban green spaces, which have a significant amount of vegetation, are allocated a value of 20 (which is a little less than 55/2). This is because there is half the amount of mineral surface in this area. Uncertainties arising from this process include a difficulty in estimating exactly how much built-up area and mineral surfaces each group is made up of. Even if the values are allocated empirically, they still reflect the trend followed by these areas composed of artificial surfaces as far as heat is concerned. The complex zones are the most difficult areas to quantify because all of the elements making up these zones are heterogenous. The value of –10 allocated to this type of zone underlines this fact.

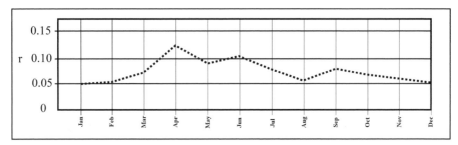

Figure 2.13. *Correlation coefficients between quantitative index values calculated from the CLC and the average temperature for each month of the year*

The final phase involves replacing the qualitative value, which shows that a climatological station belongs to one particular CLC land cover type, with an index quantitative value, similar to those developed in the previous paragraph. We now have a model that can be used to highlight the importance of vegetation in space. This new layer of information, which is understood to be the substitute to the NDVI, can be integrated and used in correlation analyses with the aim of estimating how temperature functions in space. Figure 2.13 shows the result of such an analysis that was carried out in France.

The coefficient value is low and this can be explained by several reasons:

– The vegetation index value is produced by CLC, a large-scale representation which is not perfect. Is the CLC really adapted to the issue to be resolved? It is possible that the spatial temperature variation (determined by land cover) is carried out on a small scale; a scale that is much smaller than the one provided by the CLC.

– The statistics can also be affected by conflicting constraints, that are spatial in nature. France is a large country and can have several biogeographic systems. Each biogeographic system functions independently, if not autonomously. The processes that lead to the spatial variation of the different climatic factors that are to be analyzed (such as temperature) can vary from one region to another. These differences can be found in independent spatial factors such as the estimation of biomass. These spatial factors are different spatially, but statistically they produce a general model that is quite confusing.

A certain coherent temporal variation of the index is observed. The maximum value occurs at the end of spring and at the start of summer, and the minimum values (approximately 0.05) occur in winter and (quite oddly) in August. In order to clarify the quality of this index, the values for France were compared with the NDVI for the French region of Franche-Comté.

2.4.2.4. *Validating the estimated vegetation index value*

The French region of Franche-Comté is one of the regions in France for which we have information relating to both its NDVI (calculated from satellite channels within SPOT) and to its estimated quantitative index value, which is worked out with the help of CLC. The fact that there are two sets of data makes it possible to compare the results provided by each information source, and to validate (or not) the estimation that is made by CLC. The NDVI explains the spatial temperature variation reasonably well, particularly in summer when the index reaches a value of –0.27 (which is represented by the dashed line in Figure 2.14). The correlation coefficients, which are induced by the estimated index (created by CLC), and which are represented by the full line in Figure 2.14, vary coherently with the NDVI. The change in value of the coefficients from negative to positive can be easily explained as follows: a significant amount of biomass that is present in an area has a high NDVI (210-220) and a low estimated value that is calculated by CLC (–135). The reverse situation is true for areas in which there is a low level of biomass, for example in areas that are composed of artificial surfaces.

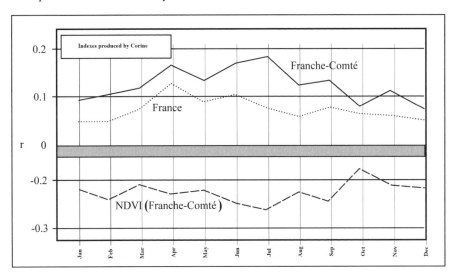

Figure 2.14. *Correlation coefficients between CLC index values and the average temperature for each month of year*

It should be noted that the weak winter coefficients have values that are in contrast to the values of the summer coefficients, and this is particularly true for July. With CLC the winter coefficients are much weaker. This fact shows that the CLC does not take into consideration any of the minute variations that exist within the different types of land cover. Figure 2.14 clearly shows that the index calculated from CLC leads to the creation of an r-level which is much higher when the correlation concerns the region of Franche-Comté (continuous line) rather than for the whole of France (dotted line). Here we can see that vegetation has various effects on temperature according to the different regions of France.

To sum up, CLC can reliably estimate the biomass that is calculated from the land cover layer that it produces. However, CLC is not as good a tool as the NDVI, and as a result, the NDVI should be used whenever possible. The satellite measurement produced by the NDVI is much more exact and is more geometrically precise than that produced by CLC.

2.4.3. *Choice of independent variables*

The majority of GIS software offers a reasonable number of functionalities, which make it possible to create a varied selection of information layers. There is the temptation to create and store a large number of these information layers and then to correlate these different layers with temperatures to try to explain the spatial variation of these temperatures in question. This approach is commendable but is not without risks. These risks will be shown in this section with the help of an example, by using distance as the explanatory variable.

Calculating a Euclidean distance on a raster image is quite a simple process, and insofar as distance is considered as one of the founding concepts of geography, there are very few geographers who can resist carrying out such calculations.

The world of climatology is composed of a large number of areas that have different characteristics (thermal characteristics, in particular) and which influence their neighboring areas. For example, there are the glaciers, whose katabatic winds transfer, by advection, masses of glacial air to their surrounding areas. Another example includes the oceans, whose maritime influences affect continental climates over a distance of several thousands of kilometers. Incorporating this notion of distance into a determinist approach, such as interpolation, stems from the hypothesis that if a geographic object has a slight influence on temperature, then the impact of this influence will decrease as distance increases. There are two key issues (amongst others) raised that need to be examined in order to validate this hypothesis and the two issues include: what really happens and how do the different variables function spatially?

The second of these issues was tested on the 1,530 French climatological stations which make up the Météo-France temperature network. The test consists of estimating temperature variation by the distance to the center of a town or city (dist-town). Distance to town is calculated from the barycenter of the "dense town" CLC polygon (CLC code 1). The hypothesis is first verified because the correlation coefficients are of a significant value for each of the three climate variables that are tested (Table 2.2): temperature decreases as the distance from large towns and cities increases, this phenomenon is also known as the "urban heat island". This fact then makes it possible to predict that the distance from a town can be used as a good estimator as far as the spatial variation of temperature is concerned. With this in mind, this information could then be stored in the GIS and used as a prediction tool within the regression models.

	AUGUST		FEBRUARY	
	average temperature	number of days >30°C	average temperature	number of days <–5°C
r	–0.31	–0.22	–0.24	0.27

Table 2.2. *Correlation coefficients (r) between three climate variables and the distance from the nearest town; 1,530 French climatological stations*

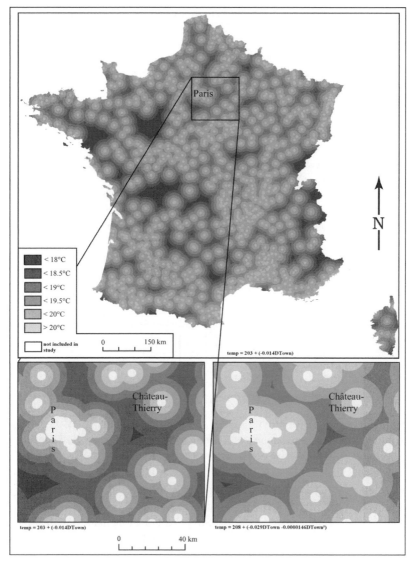

Figure 2.15. *A model showing August temperatures in relation to distance from town*

Figure 2.15 shows the result obtained after modeling: temp_Aug = f(dist-town), where August temperatures (temp_Aug) are estimated by analyzing the distance to town, which can be represented by the following equation: temp_Aug = 203 + (–0.014dist-town).

Figure 2.15 confirms what was expected: temperature is maximum (>20°C) in areas that are recognized by CLC as being dense developed towns, and then decreases the further towards the area's periphery where temperature reaches its minimum value of <18°C. In other words, temperature is at its lowest in areas that are furthest away from a town center or in areas that are located between two large towns.

Nevertheless, there is one main problem with this map: the variable "distance to town" is first considered as explanatory during the analysis phase. During the modeling phase, however, this independent variable then becomes the cause of the variations in temperature that exist. For example, from the map we find that there are rather small urban areas such as Château-Thierry, which has an exacerbated influence on its surrounding area as far as temperature is concerned.

The quadratic function equation changes this trend slightly; it reduces the influence that a town center can have on temperature. The more we expect large cities to have their own distinct heat islands, the more surprising it would be for small villages to have the same influence on temperature. The validity of the relationship that exists between these two variables needs to be confirmed to be used as a reliable model.

In order to make this point more understandable we have reproduced correlation calculations for only a select number of the largest cities in France.

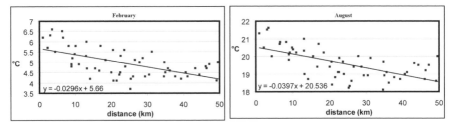

Figure 2.16. *Temperature variation for February and August in the Paris region. Variation temperature is calculated in relation to the variable "distance from the town center"*

Paris, for example, is made up of 122.5 km² dense buildings (which is represented by CLC code 1). The polygon's barycenter is located near the "Ile de la Cité", which is at the heart of Paris. Fifty-six climatological stations within a distance of 50 km from the barycenter were chosen for the study. The correlation between the temperature average for February and August (the dependent variables), and the distance of the stations to the center of Paris (the independent variable) is −0.59 and −0.63, respectively for the 2 months in question. A representation of this information can be seen in Figure 2.16.

As far as the city of Paris is concerned, there is no doubt that temperature decreases as the distance to the city center increases, and this is true up to a distance of at least 37 km from the center. This distance of ≥37 km corresponds to the vertex, which is a disguise for the variogram.

What happens at the other major cities in France that are smaller than Paris? The same test was applied to four other major cities, and in their surrounding areas (at least 50 km from their respective centers) where there was a sufficient number of climatological stations present to allow the tests to be carried out.

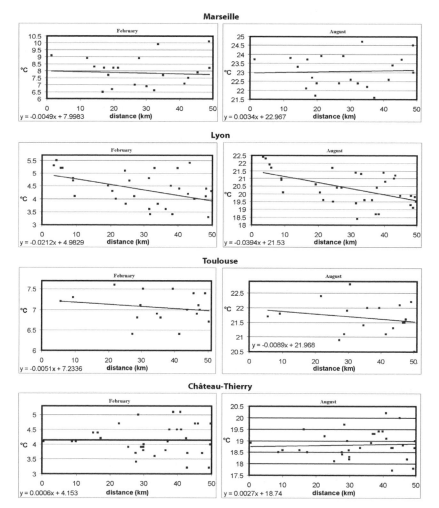

Figure 2.17. *Temperature variation around four major cities in France for Febraury and August according to the variable "distance from the town center"*

The coefficients generated for these four cities were much lower in comparison to the coefficients produced for Paris, which means that the urban island is more powerful in Paris than anywhere else in France. The city of Lyon was the only city that produced quite high values.

In the other three major cities the spatial temperature variation is only slightly dependent on, or did not dependent at all on, the distance to the town center. Nice and Bordeaux, which are not dealt with here, produced the same results. Regarding Château-Thierry, the inverse happens, i.e. the temperatures in the periphery are higher than the temperatures in the town center.

Figure 2.17 confirms and represents all of the values that can be seen in Table 2.3 in the form of a graph. The different spatial temperature variations show that distance to the town center is not the only variable that dictates this change in temperature.

	Paris	Lyon	Marseille	Toulouse	Château-Thierry
r February	−0.59	−0.45	−0.05	−0.18	0
r August	−0.63	−0.55	+0.05	−0.21	+0.06

Table 2.3. *Correlation coefficients for temperature/distance from the town center taken from four major cities in France for the months of February and August*

As more knowledge is available on how land cover influences temperature, it would be best to analyze how land cover affects the spatial variation in temperature in addition to how distance from the town center affects it.

2.4.3.1. *Vegetation index*

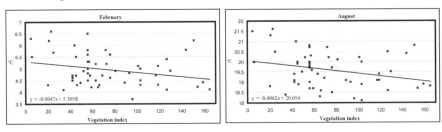

Figure 2.18. *Temperature variation in the region of Paris for February, according to the variable vegetation index*

The variable land cover is then introduced into correlation calculations by using the vegetation index estimation (see section 2.3.2.3). The index used corresponds to the average of 9 pixels taken from a 3×3 window located at each climatological station. This variable, which explains the average physical state of the surface that surrounds each station, also has a strong influence on the spatial temperature variation. Weak indexes (mineral surfaces) are associated with higher temperatures rather than higher indexes (areas in which there is a significant amount of biomass). This law of spatial organization is similar to radiation models used in physics: in this example the statistical description can be interpreted as being a causal variable.

It seems that the vegetation index is made up of a spatial structure: the index is low in the center of Paris, the densest region of the city. This index, however, progressively increases towards the periphery due to the fact that green spaces become more abundant and the heavily built-up areas become increasingly sparse. The correlation coefficient between the vegetation index and distance from town increases to 0.39. The temperature variation according to the variable vegetation increase (Figure 2.18) also follows this pattern.

The cities of Paris and Lyon have positive and negative coefficients for the two correlations temperature/distance and temperature/vegetation index, respectively (see Table 2.4). The town of Château-Thierry has low coefficients for both correlations (temperature increases as the distance to the town center increases; the value of r is negative because we are actually getting closer to the city of Paris). Marseille (in particular) and Toulouse do not present the same types of data, which shows that the statistical relationships that have been highlighted in this section cannot be applied to all towns and cities. As far as Marseille is concerned, the situation is clear: the city center is located near the coast, and is cooled down by the sea breeze. This means that the urban heat island is neutralized, and the temperature in the center of Marseille is very similar to the temperature of its periphery. As for Toulouse, the positive r value for February and August shows that there is a positive relationship (unexepected and which is inexplicable) between vegetation index and temperature.

	Paris	Lyon	Marseille	Toulouse	Château-Thierry
r February	−0.37	−0.48	−0.33	+0.32	−0.11
r August	−0.39	−0.37	−0.28	+0.27	−0.14

Table 2.4. *Correlation coefficients for the variables of temperature/vegetation index recorded in four major cities in France for February and August*

It can now be seen that other factors, such as distance to town and vegetation index, also have an influence on the spatial variation of temperature. A more detailed analysis would be able to prove this. However, the objective of our study is to find out whether the variable distance to town can be used as an independent variable to interpolate. Analyses have shown that the temperature variation in relation to distance to town is only true with a large distance for certain cities, i.e. it is true for the largest cities and not for the smaller ones. The overall negative variation between temperature and distance is due to two facts:

– the high number of climatological stations around the larger cities, and Paris in particular, takes an artificial reading from a spatial structure that is influenced by the urban island effect;

– the variable distance to town is masked by another, more causal variables that associate temperature with the land cover.

The effect that distance to the town center has on temperature has been clearly highlighted for the region of Paris. This effect exists for other major cities in France, but to a lesser extent when compared to Paris, because the density of the built-up areas in these other cities is lower than is the case for Paris. In addition, the density of the climatological stations located near the town centers of the other large cities does not make it possible to provide an accurate measurement of this variable. It is, in fact, the vegetation index that should be kept and used as the estimation variable in the regression and interpolation models. The distance from town variable, could be kept if the same level of heat as that generated in the town center is transferred by advection to the town's periphery. However, this is not the case.

2.4.4. *Spatial and scale variables*

Spatial resolution, which is fixed as soon as all data has been collected and which is uniform for all of the information layers that are held within the GIS, is another aspect that needs to be considered. Image resolution is normally chosen depending on the extent of the area to be analyzed. Depending on the actual scale of interpolation to be used, a resolution of 100, 250, and 500 m is used on a national level, whereas a resolution of 25 or 50 m is used on a regional level. As far as the study of micro-climates is concerned, this resolution may be as small as 1 or 5 m [JOL 02]. This means that, theoretically, it is possible to use different analysis and interpolation scales.

In addition, for studies that are to be carried out on only one area, France for example, several DEMs are available and they include DEMs that have resolutions of 1,000, 500, 250, 100 and 50 m. It would be a major error to confuse resolution with scale level. Different resolutions for the same piece of information provide us with an interesting method that can be used to better understand scale levels (topographic forms that are produced by a DEM with a resolution of 50 m are part of a larger scale level than those produced by a DEM with a resolution of 1 km).

To make this information easier to understand, the examples of the urban area of Dijon and the region of Franche-Comté will be used. These are two neighboring areas for which a DEM with a resolution of 50 m is available (Figure 2.19). The objective here is to examine the response generated by each resolution that estimates different climate variables [JOL 07]. This will be shown by using the example of the spatial variation of rainfall amounts.

2.4.4.1. *Available data*

The 50 m resolution DEM is available at the French National Institute for Geography. New data layers that characterize the area surrounding each climatological station are derived by computing, as has already been explained in section 2.3. Three of these layers will be tested as factors that describe the spatial variation of rainfall amount: altitude, slope, and topographic roughness.

The initial DEM is reduced 2, 5, 10, 20, 50 and 100 times with the aim of creating six other MNT that have resolutions of 100, 250, 500, 1,000, 2,500 and 5,000 m, respectively. The three data layers (altitude, slope, and roughness) are also derived from these six new DEM. There will be, at least, three layers × seven DEMs with a distinct resolution, i.e. there will be 21 independent variables to be tested in future correlations (Figure 2.20).

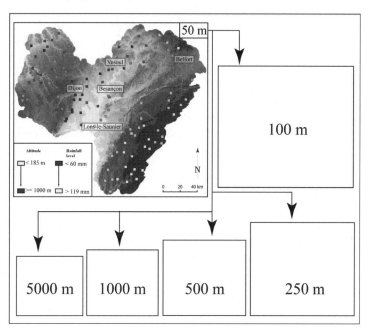

Figure 2.19. *DEM for the urban area of Dijon and for the region of Région Franche-Comté; location of climatological stations (rainfall amounts for April; 1971-2000 averages). Data source: the French National Institute for Geography, Météo-France*

One hundred and forty-three of the 3,165 climatological stations making up the Météo-France network are located in the area studied (Figure 2.19). The dependent variable, rainfall amount, is broken down into different time periods:

– the normal 1971-2000. April data was used because this month has a relatively wide range of precipitation (the minimum rainfall amount of 51 mm was recorded in the Saône plain, and the maximum rainfall amount of 146 mm was recorded in the mountain chain known as the Jura). Recordings taken from other months will also be used in order show the extent of how they differ from the April coefficients;

– the second set of data that will be used corresponds to monthly rainfall amounts, which were accumulated during 1986;

– data relating to heavy rainfall amounts that occurred during the month of April 1986 will also be used.

The spatial variation of these three sets of data (monthly averages between 1971 and 2000, monthly averages for 1986, and days of rainfall in April 1986) will be explained by the three independent variables of altitude, slope, and topographic roughness.

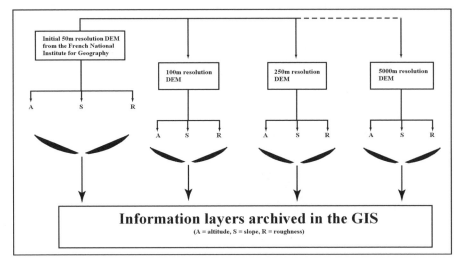

Figure 2.20. *50, 100, 250, and 5,000 m resolution derived DEM and derived information layers*

2.4.4.2. Estimating total monthly rainfall (1971-2000 normal)

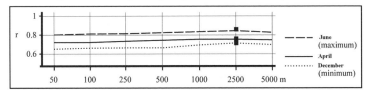

Figure 2.21. *Correlation coefficients for altitude/rainfall amount (monthly averages); the altitudes are generated by seven DEMs, each with a different resolution*

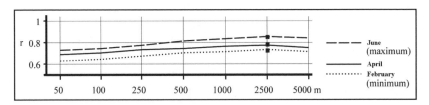

Figure 2.22. *Correlation coefficients for slope/rainfall amount (monthly averages); the slopes are generated by seven DEMs, each with a different resolution*

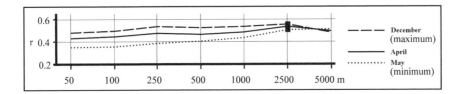

Figure 2.23. *Correlation coefficients for topographic roughness/rainfall amount (monthly averages); the roughness is generated by seven DEMs, each with a different resolution*

Altitude

The correlations between rainfall amount in April and altitude, which are considered as dependent and independent variables, respectively, measure the intensity and the direction of the vertical precipitation gradient. The first important point is that the correlations are positive: rainfall level increases with altitude. The second important point is that the coefficients very little change, regardless of the DEM which is used, and thus regardless of its resolution. This second point highlights the fact that altitude, which has a strong influence on rainfall and precipitation in general (all of the r coefficients are greater than 0.63 and reach up to 0.85), does not cause any scale effect. The altitude of the climatological stations is slightly modified when the resolutions of the DEM are changed, but this has no effect as far as the correlations are concerned [JOL 07]. The same is true for the coefficients for December (the month which produced the minimum coefficients) and June (the month which produced the maximum coefficients). Results from calculations show that altitude has a uniform influence on the precipitation gradient; irrespective of the DEM resolution the correlations remained stable for all seven resolutions, and this was true for each month of the year.

Slope and topographic roughness

Once again the correlation for these two variables is positive. The rainfall amount is much higher for steeper slopes and when the level of topographic roughness is also increased. The correlations of the two independent variables show an analog variation depending on the particular DEM in question: the relationship between slope and roughness, and rainfall amount improves as the resolution of the DEM gets larger, up to an optimal resolution of 2,500 m (Figures 2.22 and 2.23). This trend highlights the fact that the monthly rainfall averages are mainly induced by small-scale topographic factors that are present in a particular environment.

The slopes and roughness levels, which were calculated on a 50 m resolution, are related to objects whose minimum size is approximately 100 m. Contrary to this, slopes and roughness levels, which were calculated on large-scale DEMs (2,500 and 5,000 m), identified objects that were bigger than 7 km. The second set of small-scale objects has a stronger influence on the spatialization of rainfall than the first set of large-scale objects. No real change is made to this pattern, irrespective of whether the coefficients relate to April, June (maximum coefficients) or February (minimum coefficients).

2.4.4.3. Estimating the total monthly rainfall for 1986

Altitude

The correlation coefficients of altitude/rainfall amount for January, April, and November 1986 (Figure 2.24) vary slightly when different DEM resolutions are used (r values range from 0.69 to 0.72). The coefficients are quite similar to the coefficients that describe the 1971-2000 normal (Figure 2.21). November is the month that has the lowest coefficient (approximately 0.6), whilst the maximum coefficient was recorded for January 1986 (0.78). This variation pattern from one resolution to another remains: only the value of the coefficient experiences large differences from one month to the next, which means that the independent variable (causal variable) of altitude is not the same for the entire year. It would be interesting to carry out further investigations into this climatological variable.

Slope and topographic roughness

Slope (Figure 2.25) and topographic roughness (Figure 2.26), which are calculated by large-scale DEM (2,500 or 5,000 m), better explains the total monthly rainfall for April (full line in Figure 2.24) than when the same two variables are produced by lower resolution DEMs of 50 or 100 m. The difference between the two monthly extremes (January and February) is quite high, as far as the coefficients for both slope and roughness are concerned. Topographic roughness, which is calculated by intermediary resolutions (500 or 100 m), better explains the total rainfall for December (dashed line in Figure 2.26) than when the results are produced by low or high-level resolutions.

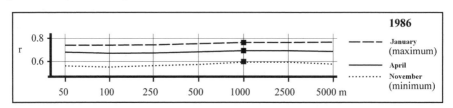

Figure 2.24. *Correlation coefficients for altitude/rainfall amount (April 1986); the altitudes are generated by seven DEMs, which have a resolution that varies from 50 to 5,000 m*

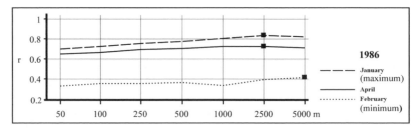

Figure 2.25. *Correlation coefficients for slope/rainfall amount (April 1986)*

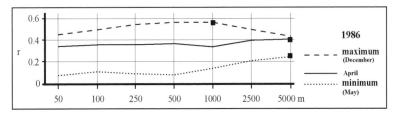

Figure 2.26. *Correlation coefficients for topographic roughness/rainfall amount (April 1986)*

2.4.4.4. *Estimating total rainfall recorded over the space of three days in April 1986*

The results of the 3 days in April 1986 for which total rainfall was recorded are shown in Figures 2.27 and 2.28. The 3 days in question were chosen in relation to the contrasting coefficients that they had, and because of the heavy rainfall that occurred on these days and which was recorded by the climatological stations located in the area being studied.

Altitude

On April 5 and 12, 1986, the level of rainfall increased as the level of altitude increased, however, the opposite was true for April 2, 1986. The negative value of r which was recorded for April 2 can be explained by an unusual abundance of rainfall in the Bourgogne region, and in particular, along the Saône plain, where altitude levels are low. The relationship between rainfall and altitude was at its strongest on April 5, 1986 (the value of r was approximately 0.75), and was lower for the other two dates.

For each of the 3 days the resolution of the DEM did not affect the coefficients. This was true for all resolutions up to 1,000 m. Resolutions above 1,000 m either slightly improved (April 2 and 5) or slightly decreased the values of the coefficients (as was the case on April 12).

Slope and topographic roughness

The values of the coefficients were positive for April 5 and 12, and negative for April 2 (Figure 2.28). The maximum coefficient value obtained by the different resolutions depended on the specific days in question:

– 5,000 m (for topographic roughness) or 500 m (for slopes) on April 5 (dashed lines);

– 50 m on April 12 (dotted lines);

– 5,000 m (for topographic roughness) or 1,000 m (for slope) on April 2 (full lines).

This was also true for the variable of altitude: in this case the coefficients also varied greatly depending on the specific day in question.

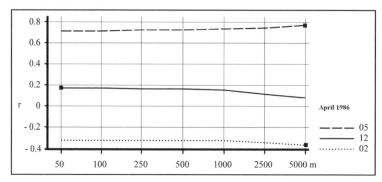

Figure 2.27. *Correlation coefficients for altitude/rainfall amount (measurements taken on April 2, 5, and 12, 1986); the altitudes were generated by seven DEMs, which had resolutions varying from 50 to 5,000 m*

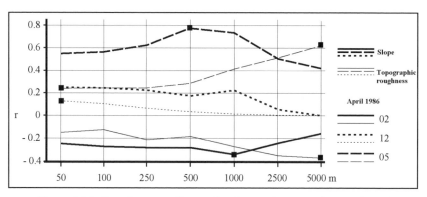

Figure 2.28. *Correlation coefficients for slope and topographic roughness/ rainfall amount (measurements taken on April 2, 5, and 12, 1986)*

Topographic variable	1971-2000 average		Monthly totals for 1986	
	Minimum	Maximum	Minimum	Maximum
Altitude	0.7	0.82	0.57	0.78
Slope	0.62	0.85	0.29	0.82
Roughness	0.35	0.55	0.08	0.57

Table 2.5. *Minimum and maximum r coefficients obtained for three different topographic variables, which are correlated with total monthly rainfall (average values and values for the year 1986)*

Coefficients resulting from correlations that relate to monthly totals (expressed over a 30 year average or throughout the year 1986) show that there is a strong relationship between the coefficients and the resolutions of the DEMs: the maximum values of the coefficients are always produced with the higher resolutions irrespective of the month of the year in question. The main difference that exists between the correlations that relate to the 30 year normal and to the individual months of 1986 is the differences in the monthly coefficients, as can be seen in Table 2.5.

However, the coefficients that relate to daily rainfall totals are more unstable. These coefficients can change from being positive one day to negative the next. For example, the maximum value produced by the coefficients may be produced by a 50 m resolution DEM or a 5,000 m resolution DEM; it depends on the day in question. The differences that exist between the coefficients are increased.

These differences result from the different classes of rainfall amounts that exist. Monthly rainfall amounts, especially those amounts that correspond to 30 year normals, will not take any daily event into consideration. Only the strong trends are taken into consideration as these trends can be easily explained by the topographic variables that are calculated by large-scale DEM and which describe the environment on a small scale. Conversely, daily rainfall totals are more erratic and spontaneous: a rain shower may only affect one small sector, and may not affect any of the neighboring areas. Precipitation, which is associated with the arrival of poor weather, is also affected by spatial irregularities. In these conditions, it is not uncommon to see that large-scale topographic variables, such as slope and roughness, influence the distribution of daily rainfall. This distribution of daily rainfall is recorded by fine-resolution DEM (with a resolution of 50 m). This fact once again highlights that there is a relationship between the spatial and temporal scales [CAR 03].

The fact that the coefficients vary with the different resolutions of the DEM is quite interesting because it shows that the scale factors (which are recorded by the different resolutions that can be applied from one DEM to another) shape the spatial organization and distribution of precipitation. In these conditions it seems that the processes of interpolation need to consider the fact that coefficients vary with resolution in order to improve the quality in estimation. Certain methods of interpolation are based on this principle [JOL 94; JOL 03]. Candidate independent variables are broken down into several windows, which are equivalent to multiple resolution DEMs, and an additional sorting process makes it possible to identify the best estimation variables and optimal window. The rain amount (or any other dependent variable as temperature) is estimated by using a multiple regression method.

2.5. Conclusion

This chapter will be summarized briefly. It is important to mention the importance of the quality of the spatial information that is available and which, to a large extent, conditions the quality of the models that are then created. The famous aphorism "garbage in, garbage out" can be applied here. This quality can be associated with one of the following characteristics:

– the location of the meteorological stations;

– the density of the stations from where the different climate variables are recorded (the interpolation of rainfall amount or temperature, which are taken from a large number of measurement stations, is much easier than the interpolation of duration of sunshine);

– the geometric precision of the tools that provide information on exogenous variables (information produced by SPOT or Landsat TM is of a better quality than the information produced by CLC), and information on the spatial resolution of the cells that make up the processed images.

Another important point, is the fact that the type of information available largely shapes the choice of interpolation method to be adopted in order to optimize the results of the interpolation process. Research that is under way and, as yet, unpublished shows that the results of interpolation are identical on a global scale, regardless of the method of interpolation that is used (regression, kriging). This is only true when the information that is available is complete (e.g. information relating to rainfall and temperature). The same cannot be said for the interpolation of duration of sunshine, or for any other variable for which the information is provided by a small number of climatological stations. In this case, the regression method is not capable of solving the problem, what kriging effectively does.

The environmental characteristics of the area to be studied (such as topography or land cover) and the spatial information that models these characteristics also largely influence the choice of the interpolation method to be used. The unpublished research mentioned in the previous paragraph has revealed that temperature and precipitation in the West of France (Brittany, Normandy, Vendée), which have a regular gradient and thus quite a high autocorrelation (precipitation and temperature scored 0.6 and 0.58, respectively, on Moran's I index), were accurately predicted by the kriging method. Conversely, the region of the Provence-Alpes-Côte-d'Azur (PACA; in the Southeast of France) had quite low autocorrelation levels (precipitation and temperature scored 0.35 and 0.3, respectively, on Moran's I index). This low autocorrelation was caused by the mountainous topography which favor the emergence of relatively independent micro-climates. An interpolation method made up of two analysis phases provides the best results (measured by the standard deviation of the residual autocorrelations) [CAR 03; FUR 95; JOL 03; LHO 05]:

– the first phase involves estimating the climate variable by using the multiple regression method (the independent variables are altitude, slope, degree of topographic roughness, etc);

– the second phase involves estimating the residual autocorrelations produced during the first phase. The kriging method is not necessarily the best interpolation method used, contrary to what many authors believe. It would be the case if the level of autocorrelation of the residuals was high. However, when the autocorrelation of the residuals is weak the cubic polynomial provides the best results.

In addition to this information, the standard deviation of the residual autocorrelations, which is produced after the two analysis phases, is much lower in Western France than in the Southeast of the country. This means that geographic characteristics (such as topography, land cover, location in relation to ocean masses, etc), which are different for each region, play a key role in determining the results of interpolation. Interpolations that are applied to the whole of France lead to the creation of problems that can only be overcome by the use of a local, rather than global approach [JOL 08; JOL 09].

Through the examples that were used in this chapter it would seem that the process of interpolation is not an easy method to use. Other examples would also have reinforced this conclusion. Technically speaking, it is not difficult to recreate the continuous field of any variable; all that needs to be done is to apply the different functions of spatial analysis software to the variables in question. The real problems arise from two main areas: from the values of the information that is introduced into the interpolation models and from the validity of the methods that are used. The interpolation process is not a simple operation. It requires a certain amount of expertise, i.e. the expertise of a statistician and of a GIS specialist. These two specialists know that the results of interpolation depend on spatial information and on analysis methods.

2.6. Bibliography

[ARN 00] ARNAUD M., EMERY X., *Estimation et interpolation spatiale: méthodes déterministes et méthodes géostatistiques*, Hermès, Paris, 2000.

[BUR 86] BURROUGH P.A., *Principles of Geographical Information Systems for Land Resources Assessment*, Oxford University Press, New York, 1986.

[CAR 82] CARREGA P., Les facteurs limitant dans le Sud des Alpes Occidentales: étude géographique, PhD thesis, Nice, 1982.

[CAR 03] CARREGA P., "Le climat aux échelles fines", *International Association for Climatology*, vol. 15, p. 19-30, 2003.

[COL 00] COLLINS F.C., BOLSTAD P.V., A comparison of spatial interpolation techniques in temperature estimation, 2000. Online at: http://www.ncgia.ucsb.edu/conf/SANTA_FE_ CD-ROM/sf_papers/ collins_fred/collins.html.

[COU 99] COURAULT D., MONESTIEZ P., "Spatial interpolation of air temperature according to atmospheric circulation patterns in southeast France", *Int J Climatol*, vol. 19, pp. 365-378, 1999.

[DUB 84] DUBRULE O., "Comparing splines and kriging", *Comp. Geosci.*, vol. 10, no. 2-3, pp. 327-333, 1984.

[ECK 89] ECKSTEIN B.A., "Evaluation of splines and weighted average interpolation algorithms", *Comp. Geosci.*, vol. 15, pp. 79-94, 1989.

[FEY 95] FEYT G., MAILLOUX H., DE SAINTIGNON M.F., "SIG et information climatique", *Revue Internationale de Géomatique*, vol. 8, no. 3-4, p. 361-376, 1995.

[FLA 01] FLAHAUT B., L'autocorrélation spatiale comme outil géostatistique d'identification des concentrations spatiales des accidents de la route, 2001. Online at: http://www.cybergeo.presse. fr/modelis/flahaut/FLAHAUT1.pdf.

[FUR 94] FURY R., JOLY D., "Interpolations spatiales (à maille de 100 mètres) des températures minimales journalières", *Publications de l'Association Internationale de Climatologie*, vol. 7, p. 542-549, 1994.

[FUR 95] FURY R., JOLY D., "Interpolation spatiale à maille fine des températures journalières", *La Météorologie*, vol. 8, no 11, pp. 36-43, 1995.

[GRA 02] GRATTON Y., Le krigeage: la méthode optimale d'interpolation spatiale, 2002. Online at: www.iag.asso.fr.

[HUD 94] HUDSON G., WACKERNAGEL H., "Mapping temperature using kriging with external drift: theory and an exemple from Scotland", *Int. J. Climatol.*, vol. 14, pp. 77-91, 1994.

[JOL 94] JOLY D., BERT E., FURY R., JAQUINOT J.P., VERMOT-DESROCHES B., *Interpolation des températures à grande échelle*, *Revue Internationale de Géomatique,* vol. 4, no. 1, pp. 55-85, 1994.

[JOL 02] JOLY D., BROSSARD T., ELVEBAKK A., FURY R., NILSEN L., "Présentation d'un SIG pour l'interpolation de températures à grande échelle; application au piémont de deux glaciers (Spitsberg)", *Publications de l'Association Internationale de Climatologie*, vol. 14, pp. 287-295, 2002.

[JOL 03] JOLY D., NILSEN L., FURY R., ELVEBAKK A., BROSSARD T., "Temperature interpolation at a large scale; test on a small area in Svalbard", *Int. J. Climatol.*, vol. 23, pp. 1637-1654, 2003.

[JOL 07] JOLY D., BROSSARD T., "Contribution of environment factors to the temperature distribution at different resolution levels on the forefield of the Loven glaciers, Svalbard", *Polar Rec.,* vol. 43, no. 227, pp. 353-359, 2007.

[JOL 08] JOLY D., BROSSARD T., CARDOT H., CAVAILHES J, HILAL M., WAVRESKY P., "Interpolation par recherche d'information locale", *Climatologie*, vol. 5, pp. 27-48, 2008.

[JOL 09] JOLY D., BROSSARD T., CARDOT H., CAVAILHES J, HILAL M., WAVRESKY P., "Interpolations locales; exemple des précipitations en France", E*space géographique*, vol. 2, pp. 157-170, 2009.

[LAB 95] LABORDE J.P., "Les différentes étapes d'une cartographie automatique: exemple de la carte pluviométrique de l'Algérie du Nord", *Publications de l'Association Internationale de Climatologie*, vol. 8, pp. 37-46, 1995.

[LAN 94] LANGLOIS P., "Une transformation élastique du plan basée sur un modèle d'interaction spatiale, applications en géomatique", *Geographic Information Systems, GDR 1041 Cassini*, INSA Lyon, 13-14 Oct 1994, pp. 241-250.

[LAN 96] LANGLOIS P., "Interpolation de données ponctuelles", *Cybergeo: Eur J Geogr*, vol. 8, 1996.

[LAS 94] LASLETT G.M., "Kriging and splines: an empirical comparison of their predictive performance in some applications", *J. Am. Stat. Assoc.*, vol. 89, pp. 391-409, 1994.

[LHO 05] LHOTELLIER R., "Spatialisation des températures en zone de montagne alpine", PhD thesis, University Joseph Fourier, Grenoble 1, 2005.

[MER 01] MERLIER C., "Interpolation des données spatiales en climatologie et conception optimale des réseaux climatologiques", Annex from the Météo-France report on its involvement with the commission for climatology (CCI) de l'OMM., 2001.

[MIT 99] MITAS L., MITASOVA H, "Spatial interpolation", in P.A. Longley, M.F. Goodchild, D.J. Maguire, D.W. Rhind, *Geographical Information Systems: Principles and Technical Issues*, vol. 1, John Wiley, New-York, pp. 481-492, 1999.

[OLI 90] OLIVER M.A., "Kriging: a method of interpolation for geographical information systems", *Int. J. Geogr. Inf. Sys.*, vol. 4, no. 4, pp. 313-332, 1990.

[PHI 01] PHILIPPE A., PIÉGAY H., "Pratique de l'analyse de l'autocorrélation spatiale en géomorphologie: définitions opératoires et tests", *Phys. Quatern. Geogr.*, vol. 55, no. 2, pp. 111-129, 2001.

[TAB 05] TABSOBA D., FORTIN V., ANCTIL F., HACHÉ M., "Apport de la technique du krigeage avec dérive externe pour une cartographie raisonnée de l'équivalent en eau de la neige: application aux bassins de la rivière Gatineau", *Can. J. Civil Eng.*, vol. 32, pp. 289-297, 2005.

Chapter 3

Geographical Information, Remote Sensing and Climatology

3.1. Introduction

Research carried out in the field of climatology is based on weather observations that are recorded in a certain number of climatological stations. These observations can sometimes be recorded over a very long period of time, as is the case in China (where rainfall records date back to the Middle Ages), and in England where rainfall records date back to the 18th century. In his thesis, "Climate history since the year 1000", the historian E. Leroy Ladurie [LER 83] proved the usefulness of written or iconographic documents for the study of past climates. Different climate events have also strongly influenced history, for example, during war [PEG 89]. However, our knowledge of the Earth's climate is not only used for strategic purposes. It is, above all, the agricultural industry that depends on climatic conditions and, in some cases, the local climate can be seen as a useful resource or as a constraint. This means that climate does not play a neutral role in space or in the world of geographical reasoning: it is one of the elements that influences the land that is used by humans. Different societies function in different ways and this depends on a particular country's level of development, on a society's knowledge of different climate factors, and on the methods that it has available so that it can protect itself from harmful climatic events.

Since the 1970s satellite observation of the Earth's atmosphere has greatly transformed and improved our knowledge of how different climate mechanisms function. From a very early point in the 1970s climatologists have understood the importance and significance of satellite data for their research: this can be seen in pioneering research such as that carried out by Tonnerre-Guérin [TON 76], Mounier and Pagney [MOU 82], and the success of Barrett's manual [BAR 74]. Other pioneering work carried out in this field include the contents of *Satellites et Climatologie* the 3rd Conference held by the International Association of

---

Chapter written by Vincent DUBREUIL.

Climatology [MOU 90] as well as the many theses written in France [TAB 89; DUB 94; WAH 97] or in Brazil [LOM 89; MEN 95], where data produced by remote sensing proved to be extremely useful. More recently, the initiative created by the Research group Meteosat Second Generation of the French national center for scientific research has meant more collaboration between different geographers in order to produce and process Meteosat Second Generation images. The aim of this chapter is to analyze the use of satellites, which have greatly improved our knowledge of the spatialization of climate data over the past 30 years [KER 04]. An explanation will be given as to why satellite data plays a decisive role when it comes to increasing the density of geographical information for research that is carried out in regions in which there are not many conventional climatological stations available. For example, in Mato Grosso the rapid expansion of crops has led to an increasing demand on the climate. After an overview of the available satellite data, we will use examples of cases where geostationary data are used to estimate rainfall levels, and where SPOT-vegetation satellite data are used to estimate the extension of vegetation and crop cover.

3.2. The development phases of meteorological satellites

3.2.1. *Insufficient sources on the land*

Up until the 1960s and the 1970s, the basic data used were measurements that were taken on the ground. This involved observing and recording data taken in stations that were located on the land itself and which was subject to strict rules enforced all over the world by the World Meteorological Organization (WMO), by Météo-France in France and by the National Institute of Meteorology in Brazil. The data available are limited and extremely localized, and are treated with care over the long term. In order for different datasets to be compared, these observations and recordings need to be standardized (this is generally the case for any given country), for example, the temperature recordings need to be measured at a specific level (generally 2 m). The quality of the information available varies depending on the type of material used and on the density of the network of climatological stations.

Using different climate data can also lead to the creation of certain problems [ER30 81]. First, it is necessary to highlight the uneven quality of the observations: reading errors, as well as problems with the climatological stations, are factors that need to be considered whenever figures are analyzed. It is sometimes possible to detect such errors and even correct them by using simple statistical methods. However, one problem remains. The measuring equipment can sometimes be unsuitable for use in turbulent atmospheric conditions, for example, rain gauges tend to overflow during periods of heavy rain (this happened during a storm in Nimes, Southern France, in 1988); anemometers are sometimes blown away by strong gusts of winds (winds similar to hurricane force) and heavy storms, such as that which occurred in Western France in 1987; these are examples of extreme weather conditions that are not always considered when it comes to analyzing climate data.

In addition, other more subtle drifts have also been noticed. These drifts are associated with changes that are made to the measuring equipment or to the measurement sites themselves. Changes made to the measuring equipment or to the time of measurement affect the continuity of the series of observations: it is now possible to note a substantial difference in temperatures measured in stations that have a wooden shelter and those that have a plastic shelter. The change in equipment used to measure the duration of sunshine that occurred in France in 1990 led to a break in the series of observations. It is known that tipping bucket rain gauges underestimate rainfall levels. It is also necessary to consider any changes that have been made to the environment of a site from where measurements are taken: the continued rise in temperatures over the last several decades also comes from the fact that the measurements are now recorded in or near urban areas. There has also been a spectacular decrease in temperature and in rainfall levels from climatological stations that are located in areas where a neighboring tree has taken up too much space or where a new building has been constructed. This means that we should not read too much into the available data that initially appears to be precise. A certain amount of rigor and attention to stations metadata is required in order to study long series of measurements.

The first international network of meteorological measurements was created by Ferdinand II, the Grand Duke of Tuscany. He developed a network of 10 measurement stations, which stretched from Florence and Pisa to Paris and Warsaw. The network was in operation from 1654-1667, and was abandoned because of the hostile views that the Catholic Church held against intellectuals [ROC 93]. In France modern meteorological observation dates back to the second half of the 19th century. On November 14, 1854, in the middle of the Crimean War, a violent storm occurred and resulted in several ships from the Franco-Anglo fleet being washed ashore on the coastline of the Black Sea. Could this military disaster have been avoided? In order to answer this question the Emperor Napoleon III asked Urbain Le Verrier for some expert advice. Urbain Le Verrier was a widely recognized scientist. Le Verrier quickly showed Napoleon III that the same storm had raged across France only 3 days earlier and then progressed across the rest of Europe. Therefore, this storm could have been forecast. Some weeks later, on February 19, 1855, Le Verrier presented a weather map to the Academy of Science. On this weather map he had information about the storm and its path across Europe on the morning that it occurred. Not only did he use the fact that there was a war taking place to his advantage, but he also benefited from the recent development that was taking place, the invention of the electrical telegraph. The invention of the electrical telegraph made it possible to send information to remote places. From this date a storm warning service was created in France, and this service was the ancestor of Météo-France, which exists today. Elsewhere in Europe, numerous meteorological commissions would be created, which were the results of events that took place during World War I. During World War I, weather forecasting was often a key element in whether a particular campaign succeeded or failed. This is still true today, and was the case during the Gulf War in 1991: the correct organization of military operations (including stealth planes) depended heavily on meteorological and climate conditions. In certain countries it is difficult to access such data as it is top secret. In some aspects, weather forecasting is often seen as a product of war or seen as the 'Daughter of Mars' [PEG 89]; conversely, it is also seen as being a branch of geography that is used to make war.

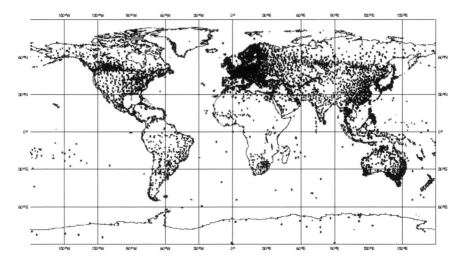

Figure 3.1. *The WMO network. Data source: WMO,*
http://www.wmo.ch/index-en.html (June 2005)

The global network map of the some 20,000 climatological stations that are officially recognized by the WMO provides geographers with lots of useful information (Figure 3.1). From the map it can be seen that there are more climatological stations located on land than on the oceans, yet the oceans make up more than 70% of the Earth's surface. As far as measurements on the oceans are concerned, there is a network of 450 drifting weather buoys, and measurements are taken on board more than 7,000 ships. It is also possible to see on the map that there is an over-representation of climatological stations in the northern hemisphere, especially in regions found at the middle latitudes. With this information in mind it could be said that meteorological observation is a luxury, and is only reserved for the richer, more economically developed countries. With this map it is also possible to identify sparsely populated areas, such as Siberia and areas along the Trans-Siberian route. One cannot help but notice that meteorological information is discontinuous. For certain information, such as the information that is produced by radiosonde exploration, the situation is much worse as there are not very many climatological stations that provide data in this way (less than 500 of these stations exist in the world; Pailleux, 2002). It is possible to find the same number of such stations in Germany as there is in the whole of Africa. There are more of these stations in the USA than can be found on the oceans. The problems that are associated with the density of the network providing information relating to rainfall (Figure 3.2) can be illustrated by using the example of the networks in Mato Grosso (which has a surface area of 907,000 km^2) and France (which has a surface area of 551,000 km^2).

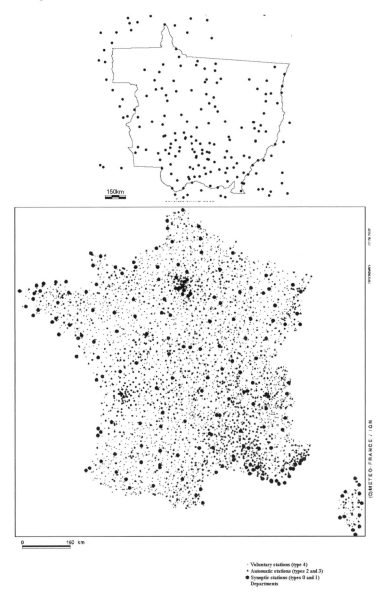

Figure 3.2. *The network of stations measuring rainfall in Mato Grosso and France. Data source: Météo-France. The two diagrams are drawn to the same scale*

The map shows that there is a big difference between northern and southern part of the state of Mato Grosso (and Bolivia), where information relating to rainfall levels has been available for more than 30 years. In northern regions, which have a latitude of 15°S, the climatological stations have only been in place since the end of the 1970s, or more recently in some cases. In these northern regions, the stations were installed at the beginning of colonial farming programs: the oldest stations in the northern regions are Vera and Porto dos Gauchos, which were installed in 1973, and Alta Floresta, which was installed in 1978. Just as Marchand [MAR 85] had noticed for Ireland, this spatial distribution summarizes the history of the development of Mato Grosso, and also shows the shift of the main area of interest from the southern part of the state to the north. At the moment, the installation of new stations depends on the movement of the population, which does not follow a precise logic.

In France, the Météo-France network (whose headquarters are located in Toulouse) relies on departmental associations: the departmental centers for meteorology. In France there are just over 3,000 measurement sites, which are spread out over the country (one site for every 180 km^2) that produce the same quality of data. Numerous spatial gaps exist, especially in mountainous areas (in high-altitude areas or on the north-facing sides of mountains). To make up for these gaps, irrespective of the region being studied, satellites are able to produce a regular mesh spatial interpolation map. Contrary to the data collected at ground level, which is often made up of a mixture of different variables, satellite information is coherent and homogenous. The frequency of their observations is much greater than the frequency of observations produced by official measurement sites. This information provides us with a better understanding as to why satellite observation techniques have adopted a somewhat operational character.

3.2.2. The progressive development of operational techniques

The observation of the Earth's atmosphere has been a priority ever since the introduction of satellite remote sensing. From the 1960s right up until the end of the 1980s the majority of satellites were used for military purposes. There are, on average, 50 satellite launches per year and approximately 80% of them are used for strategic reasons. Only 2-4 of such launches were used for civil observations (mainly by using the Landsat and SPOT programs). Over the past three decades there has been a significant number of satellite launches for meteorological purposes; on average, five to six launches per year. Since 1991 and the end of the Soviet and communist regimes in Eastern Europe, the number of meteorological and civil observation satellites launched has overtaken the number of military satellites launched. The number of military launches has been halved and there is now a wider range of countries that launch the satellites. Pioneering countries, such as the USA and Russia, are facing competition from countries such as Europe, Japan, and the emerging countries of China, Brazil, India, and Israel, which have their own space programs [VER 92]. Looking back at the history of satellite observation over the past 45 years it is possible to divide the 45 years into four successive phases, with each phase lasting for a period of 12 years.

3.2.2.1. *The development phases of meteorological remote sensing*

The first benefits associated with meteorological remote sensing date back to the end of the Second World War. In 1947 the Americans converted V2 rockets into measurement devices and optical cameras. The first photographs of cloud systems date back to this period. The cameras were taken to an altitude of 110-165 km from where the photographs were taken. Although these devices were not similar to the satellites that exist today, their observations made it possible to study and analyze the complexity of cloud systems (during the era of Norwegian frontology). The use of these devices also showed that it was important to have a global view of such systems.

The dawning of the meteorological remote sensing era really began after the launch of Sputnik (October 1957) and the creation of the National Aeronautics and Space Administration (NASA) in March 1958, after it launched its first satellite known as First Earth Radiation Experiment. The most important date in the history of meteorological remote sensing is the April 1, 1960. On this date the Television and Infrared Operating System (TIROS-1) was launched. TIROS-1 was the first real meteorological satellite that showed the complexity and importance of the satellite approach and the need for international co-operation and the development of a global approach in order to monitor the Earth's atmosphere. In December 1962, with the support of the United Nations (UN), the World Weather Watch (under the guise of the WMO) and Global Atmospheric Research Program (GARP) were created.

The 1960s saw the creation of the first generation of TIROS orbiting satellites. These were 10 similar satellites, which were in use until 1965 and orbited the Earth at an angle of 48° to 58° from the Equator, focusing on low and middle-range latitudes. The Environmental Science Services Administration (ESSA) was also created in the 1960s. The nine ESSA satellites, which were launched between 1966 and 1969, had almost polar orbits and they recorded information from the poles. All of these satellites showed the operational feasibility of the satellites' sensors, of the satellites' Automatic Picture Transmissions (APT) systems, and of their on-board storage (data that was gathered on land in the states of Virginia and Alaska in the USA). Regarding Russia, the first meteorological satellite, COSMOS-122, was launched in 1966, and in 1969 METEOR-1 was launched. METEOR-1 was the first operational satellite to be equipped with an observation TV on board.

The 1970s was a decade of development and building on the meteorological observation systems that were created in the 1960s. Despite global economic difficulties, the Apollo missions launched by NASA made it possible to develop new, and more operational programs. It should be pointed out that geostationary platforms were also developed in the 1970s, with the launch of GOES-1 in 1974, followed by METEOSAT-1 and GMS-1 in 1977. A second generation of orbiting satellites was born with the arrival of the Improved Tiros Operational System (ITOS), which was launched in January 1970. ITOS was made up of a double transmission system (APT + storage) and could take images with resolutions that ranged from 1-4 km in visible and in infra-red. In December 1970 the successor of ESSA became known as NOAA-1. NOAA refers to the National Oceanic and Atmospheric Administration. The NOAA satellites were used by the US Department of Commerce to produce images for operational and commercial purposes. This first series of NOAA satellites were equipped with a very high resolution radiometer (VHRR) and a vertical temperature sounder. In 1975, the second generation of Russian METEOR satellites was also equipped with an imaging radiometer. The first high-resolution observation satellites were also sent into orbit during the 1970s. These satellites included LANDSAT MSS in 1972, and then SPOT in 1984.

An important turning point occurred in 1978 with the arrival of a new generation of NOAA sensors. The advanced very high resolution radiometer (AVHRR), which was on board the TIROS-N that was launched in 1978, observed the Earth's surface in four different spectral bands: 0.55-0.9 µm, 0.73-1.1 µm, 3.5-3.9 µm, and 10.5-11.5 µm. From the NOAA-6, and for any subsequent NOAA satellite, the passing band of channel 1 was reduced in size so that it would be easier to see and distinguish between different cloud cover (0.55-0.7 µm then 0.58-0.68 µm). It has been possible to calculate vegetation indexes since 1980. These indexes combine information gathered in channels 1 and 2, which is close to the infrared region of the electromagnetic spectrum. From the NOAA-7 and for any subsequent satellite a fifth channel was added to the sensors (with a spectral band 11.5-12.5 µm) with the aim of improving estimations of temperatures relating to the surface of the sea [DSO 96].

In 1998, the AVHRR-3, which was installed on board the NOAA-15, had a sixth channel added to it (with a spectral band of 1.58 – 1.64 µm) to carry out research on ice and snow. In 1978 a series of experimental platforms, such as SEASAT and NIMBUS, were launched. The 1980s was the decade in which long-term operational programs were created, some of which are still functioning today. Great efforts were also made so that it could be possible to integrate satellite data into different weather-forecasting models. The first global model, which stored data relating to wind (produced by geostationary platforms on a daily basis), was created at the European Centre for Medium-Range Weather Forecasts (ECMWF) in Reading in August 1979 [PAI 02]. The focus, however, seemed to be on the development of different models, and the actual meteorological spatial technical was forced into the background. Scientists were focused on a new perspective based on global change; as a result, the benefits of remote sensing were not fully appreciated.

The 1990s were also marked by a series of problems and the arrival of a new generation of sensors. The different satellite programs experienced a wide range of problems, and these included: the explosion of the space shuttle Challenger, during take-off in 1993; difficulties associated with the development and upgrading of the GOES-next series (a three-way stabilization system); the failure of the LANDSAT 6 mission; and the loss of the NOAA-13 satellite in 1993 after being in orbit for only 13 days, etc. At the same time, however, new observation systems were being created and these included: radar images and altimetric radars (TOPEX Poseidon, ERS, etc), as well as atmospheric sounding instruments that produced better results than their predecessors (ATOVS on NOAA-15 since 1997), and instruments used to measure ocean trends, as well as instruments used to measure any interactions taking place between the ocean and the atmosphere (POLDER, AQUA, etc). Spatial (1m resolution) and spectral resolutions (several channels were being used) became more specialized during the 1990s. There was also the evolution of geostationary observations with the arrival of the new series of the American GOES, and the arrival of the Meteosat second-generation (MSG).

During this period the focus of scientists was starting to change as the series of satellite data available was starting to be used for climatological purposes, such as global change studies. This was possible because the data were taken from observations that dated back to the end of the 1970s, in other words the NOAA or GOES satellites had produced data for approximately 20 years. During the 1990s there was a significant increase in the amount of satellite data available for the Earth's atmosphere, the oceans, biomasses (both continental and marine), and the cryosphere, etc [GUR 93].

Over a period of 40 years the advances in spatial meteorological technology made it possible to completely change the way in which climatologists and geographers worked. Furthermore, a large quantity of data has been collected by a variety of different sensors, and can be easily accessed by everyone in the scientific community.

3.2.2.2. The different sensors and images available

Meteorological satellites make up the most important part of the observation satellites [BUR 91]. A distinction is made between geostationary satellites (as can be seen in Figure 3.3) and orbiting satellites (as can be seen in Figure 3.4) [RAI 03]. The use of these satellites will be explained in more detail in the next two parts of this chapter. In addition to these geostationary and orbiting satellites, numerous meteorological satellite programs have also been developed since the 1960s. The most commonly used programs include the NIMBUS satellites (which monitor the ozone and improve upon the sensors that are already in place), TOPEX-POSEIDON (altimetric radar), the Defense Meteorological Satellite Program (DMSP), which is a system of military satellites with microwave sensors used to measure precipitation, and the Tropical Rainfall Measuring Mission (TRMM). The TRMM was the first precipitation radar used on board a satellite. Europe also got involved in the space race with the launch of ENVISAT in March 2002. ENVISAT is a European satellite that monitors the environment and is equipped with MERIS (15 spectral bands found in the visible light and near infrared regions of the electromagnetic spectrum with resolution images of 300 m or 1.2 km). At the end of 2006 Europe also sent its first orbiting meteorological satellite into orbit. The satellite was known as METOP and was the equivalent of the American NOAA series. METOP was equipped with the AVHRR-3 and monitored the ozone as well as ocean-winds, etc.

Meteorological satellites have become increasingly important ever since the correct forecasting of cyclone Camille, which occurred in the southeast of the US in

1969. Camille was one of the most powerful cyclones ever to hit the US. The alert, which was raised thanks to the information provided by satellite observations, undoubtedly saved the lives of hundreds of people. The advantage of satellites is linked to the high frequency of their observations and also to the fact that they monitor the spatial dynamics of meteorological phenomena, in other words they highlight any changes that occur in the weather, good or bad [BAD 95]. Since the 1970s it has become possible to produce a spatialization of thermal data and cloud cover so that an accurate weather forecast can be made. It is also believed that these satellites have increased the length of a weak forecast period (of 5 days) by up to 2 days. Satellites used in climatology are most commonly used for the following: for monitoring cloud cover (by using the ISCCP program), for estimating precipitation levels from METEOSAT data, for monitoring drought, and for monitoring of the phenomenon known as El-Niño.

3.2.2.3. *From the satellite to the image: methods and importance*

Waves are ranked in relation to their wavelength in a vacuum. In the electromagnetic spectrum, the human eye is only able to see a limited range of wavelengths, visible waves range from 0.4-0.7 μm. Within the electromagnetic spectrum the shortest wavelengths are the ultraviolet waves and the longest wavelengths include infrared waves, radio waves, etc. All of these different wavelengths occur whenever there is a specific energy level in place. Meteorological satellites measure a fraction of the waves, which have a short wavelength, in order to find out the albedo of the surfaces. In addition, these satellites measure the geographical information necessary for analyzing the radiation that exists on a given part of the Earth's surface. The second wave type measured is the infrared waves because the frequency at which these waves are emitted depends on the temperature of a particular surface (band width 10-12 μm). Other wavelengths are important for obtaining information about the Earth's surface or atmosphere: near-infrared waves are used to monitor vegetation, medium infrared waves are used to measure absorption by atmospheric water, and microwaves are used to measure the water content of water droplets that are found in clouds, etc.

The satellite detectors (radiometers) are sensitive to the energy of the wave that they are measuring: calibrating the detectors means that it is possible to record values of certain wavelengths or a particular region of the electromagnetic spectrum. This idea of calibration defines the purpose of a channel, or it can also be used to define the spectral resolution of a sensor. Depending on the distance from the satellite to the surface of the Earth and how often the data is collected by the satellite's radiometer, the distance for any two successive measurements will be different: this is known as the sensor's primary spatial resolution, and it is here that the pixel (picture element) is defined. The pixel is the smallest element of a satellite image and corresponds to a radiometric measurement. For example, the satellite SPOT-4 has a spatial resolution of 10 m in panchromatic mode, and 20 m in multiband mode. The LANDSAT-TM satellite has a resolution of 30 m, whereas the METEOSAT-3 satellite has a resolution that ranges from 2.5-10 km.

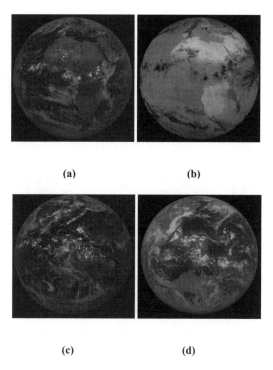

(a)   (b)

(c)   (d)

Figure 3.3. *Some examples of images produced by geostationary satellites: a) an MSG image, taken on the July 27, 2006 at 12:00 GMT using visible band. The very reflective surfaces (clouds) can be seen in white; the surfaces that absorb the short radiations (oceans) are the dark areas. b) An MSG image, taken on the July 27, 2006 at 12:00 GMT using infrared band. The cold surfaces are the dark areas, and the warm surfaces are the bright areas. c) A GOES-12 image, taken on the July 27, 2006 at 18:00 GMT using visible band: same description as image a). d) A MTSAT image, taken on the July 27, 2006 at 03:00 GMT using visible band: same description as image a)*

The result of all of these different aspects when they are linked together is the creation of a digital image (a simple matrix of points which have a value between 0 and 255) in grayscale. The word photograph is not used as no photographic film is ever printed. With the help of technology, color is then added to the image. The color allocated is done so by using the operation false color. This process can sometimes be carried out on just one channel a color is allocated to a range of pixel values. This technique is commonly used for distinguishing between the different temperatures of different surfaces. It is also possible to superimpose several channels in the three basic color channels (red, green, and blue). This process is known as color composition and it is commonly used to distinguish between different cloud types.

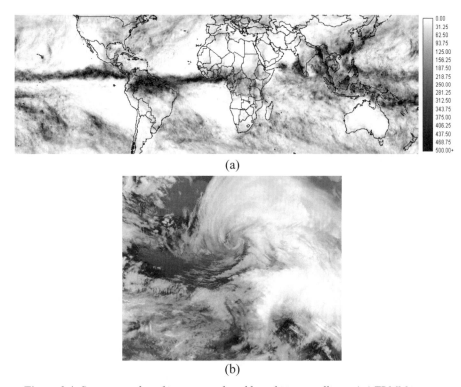

(a)

(b)

Figure 3.4. *Some examples of images produced by orbiting satellites. a) A TRMM image: monthly analysis of predicted rainfall (in millimeters) for the month of June 2006. Spatial resolution: 0.25°. b) A NOAA-15 AVHRR image from the December 26, 1999 at 08:00 GMT in thermal infrared (channel 4). The cold surfaces are the areas in white and the warm surfaces are the areas in black. Resolution: 1.1 km*

The use of satellites increases the quality of forecasting on a global level. Maps that show the location of climatological stations all over the world (only one-tenth of the entire world is well-known with this global network) are also used to provide information about inhospitable areas and about regions where it is very difficult or sometimes dangerous to obtain information. These areas include oceans, deserts, high latitudes and altitudes, and areas in which there are lots of cyclones. Satellites produce data over a short period of time and at a low cost. The frequency of the observations carried out is important for meteorological phenomena that last for a short period of time (a few hours for a storm, a few days for a depression), and for phenomena that occur over a wide surface area. The data collected by the satellites are much more homogenous than the data collected on the ground because there is no frontier effect. The data can be used in both digital and spatialized formats. The frequency of radiometer scanning leads to the creation of a grid of measurement points that are evenly distributed over the Earth's surface and which can be easily used in GIS software. The fact that there are different measurement points makes it easier to create a map of the different climate variables. However, it would be a mistake to think that this type of data could replace on-site measurements, because satellites do not measure the same variables, and this type of data needs to be calibrated and validated by conventional data. With this in mind, the data produced by satellites are seen as providing information that complements the information used for the spatial representation of the climate variables. This point will be explained through the use of examples chosen for the following sections of this chapter.

3.3. Examples of how geostationary data are used in Brazil

3.3.1. *Geostationary satellites: GOES and Meteosat*

Geostationary satellites have been monitoring meteorological activity in most of the world since 1977 (excluding areas of high altitude). A group of five satellites are responsible for regularly monitoring meteorological activity on a world scale. The Meteosat satellites (created by the European Space Agency) monitor activity that takes place in the Europe-Africa zone. The American GOES satellites monitor activity that occurs in the America-Atlantic and America-Pacific zones. The Japanese GMS satellites monitor activity in the Asia-Pacific-Australasia zone, and the Asia-Indian Ocean zone is monitored by Russian satellites (GOMS), Indian satellites (INSAT) or Chinese satellites (FY-2). These are geostationary satellites. The term geostationary is used because the satellites seem to be stationary in relation to the surface of the Earth. What actually happens is that the satellites move at the same speed as the Earth, but at an altitude of 36,000 km. Despite having a spatial resolution of between 5 and 10 km for a long time, as well as recording data every 30 minutes or every hour for over 10 years now, geostationary satellites have improved dramatically over the last decade. Since 2002, METEOSAT entered into a new era, with the launch of MSG. Performance has improved with an increase in spatial resolution in the visible band (1 km as was the case for the American GOES satellites in 1994). Performance has also improved thanks to the fact that an image is captured every 15 minutes and there are new spectral bands that can be used to produce the images (see Table 3.1).

The high number of images taken makes it possible to record data on a daily bases (as used for TV news headlines), on a monthly basis, or on a yearly basis. These images are the key elements to modern weather forecasting because with these images it is possible to recreate movements that take place in the atmosphere (or wind field) from the continuous observation of cloud formations. Clouds are the first elements of the Earth's atmosphere that are observed by satellites. The International Satellite Cloud Climatology Program (ISCCP) was created in 1983 [SCH 83]. The ISCCP was used to create a global database so that it would be possibly to study clouds from data that was provided by the geostationary and orbiting satellites. The data was analyzed on a global scale every 3 hours and at different resolutions. The resolutions ranged from 30-250 km. Monthly data analyses are also available. On a local and regional scale, however, several different research studies have used satellite data to monitor cloud cover [MOU 81]. We will use the example of Mato Grosso in order to highlight the link that exists between rainfall data on the ground and satellite signals. The Northeast Region ("Nordeste") in Brazil will also be studied in order to study sea-breeze fronts.

	Meteosat	**MSG**
Visible	one channel (5 km)	4 channels with a High Resolution Visible of 1 km
Medium infrared	one channel (10 km)	3 channels (3 km)
Thermal infrared	one channel (10 km)	5 channels (3 km)
Frequency of images	30 minutes	15 minutes

Table 3.1. *A comparison of performances produced by Meteosat and MSG*

3.3.2. *Monitoring sea-breeze fronts in the northeast region of Brazil*

Coastlines are the areas where specific climatic and meteorological conditions develop, and in particular, the development of mesoscale patterns, such as sea breezes. Such mesoscale patterns lead to the creation of weather phenomena that only occur along the coast or in coastal areas that have an extensive land border. These areas tend to be relatively fresh, humid, and sunny in comparison to the interior regions of the country. This type of weather also has an influence on human activity, and influences activity such as tourism, atmospheric pollution, etc. The sea breeze that develops along the coastline is marked by a change in the wind direction on the surface of the land that runs along the specific coastline in question. In addition to a change in the wind direction, the wind's speed also increases and there is a decrease in temperature as well as an increase in relative humidity and a change in cloud cover. Sea air is denser, more stable and fresher than continental air, and due to these three facts the sea air is able to raise the warmer, more instable continental air, which, in turn, speeds up the rate of convection of the sea-breeze front, and as a result, a line of cumulus type clouds are formed. This line of clouds is formed parallel to the coast and is then pushed further inland by the sea breeze. Thanks to this type of cloud cover created further inland, it is possible to clearly see the sea-breeze front on satellite images as the coastline and the land that runs along the coast benefit from a clear sky [PLA 06]. Remote sensing has often been used to demonstrate such cloud activity associated with sea breezes, however, no systematic analysis of this phenomenon has ever taken place.

In the example, visible images from the GOES-8 satellite were used (they have wavelengths which range between 0.58 and 0.68 μm). These images are used because recordings that are made within this range of wavelengths means that it is possible to distinguish between different cloud types in relation to their albedo: the albedo of the cloud types varies in relation to a cloud's thickness and its density. The images created are used to identify the extent and edges of the clouds, as well as to identify cloud formations. The images are also used to distinguish between low-lying cloud, which is present on the land or on the sea. For the purposes of this study the data used were collected everyday at 18:00 UTC (Coordinated Universal Time, which corresponded to 15:00 local time in the northeast of Brazil for the months of September to December 2000). The method involved identifying cloud cover that was linked to sea-breeze fronts (lines of cumulus clouds on land parallel to the coastline), and this was followed by plotting the sea-breeze fronts on a map in order to estimate how often they were created.

In the northeast region of Brazil, the location of the sea-breeze fronts shows that there is a very steep spatial gradient. The uneven topography of the land exposed to the East coast (between Salvador and Recife) makes it difficult to identify sea-breeze fronts because they tend to be hidden by other cloud formations that are influenced by mountains and orographic effects (Figure 3.5). At 18:00 UTC, the sea-breeze fronts reach further inland (known as penetration) as the dry season advances: in September it is common for sea-breeze fronts to be recorded further inland at a distance of between 40 and 80 km from the coast, with the average distance being approximately 60 km. The maximum distance from the coast of between 60 and 100 km was recorded in November 2000. In French Guiana and in the areas surrounding the mouths of the Amazon River, this distance is at least 100 km. When the sea-breeze front penetrates so far inland this means that the cloud cover associated with the sea-front breeze starts to mix with cloud cover associated with the convection currents that occur above forest areas. The slightly indented nature of the Ceara coastline makes it easier to identify sea-breeze fronts (Figure 4.5) and also makes it possible to determine how the sea-breeze fronts adapt depending on how far inland they have penetrated. The strongest, most developed sea-breeze fronts are associated with weak synoptic winds blowing in the opposite direction of the sea-breeze. A weak synoptic wind that blows in the same direction as the sea breeze leads to the sea-breeze front moving further and further inland.

Figure 3.5. *A GOES colored composition from the September 8, 2006, 18:00 GMT (see color section)*

By focusing on the thermal gradient of the earth-sea surface during the dry season it has become possible to create thermal conditions that favor the development of powerful sea-breezes, which can penetrate quite far inland. The trade winds carry the atmospheric humidity necessary for the formation of clouds that are associated with the sea-breezes. It seems that from a threshold temperature of 4°C, a change in temperature of 1°C is equivalent to the sea-breeze front moving further inland by a distance of 10 km. This approach, which still needs to be automated and systematized [COR 06], shows that, in addition to the ISCCP approach, the geostationary imaging has potential when it comes to the spatio-temporal monitoring of regional cloud conditions that are associated with a particular type of weather.

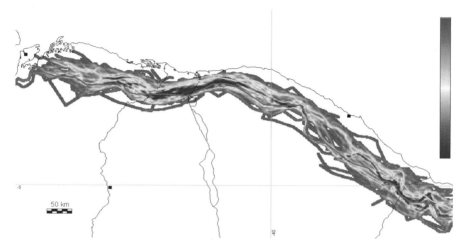

Figure 3.6. *An analysis of the recordings of sea-breeze fronts taken between September and December 2000. The low frequencies are shown in green, and the high frequencies are shown in brown. The blue line shows the average location of the sea-breeze front at 18:00 for the month of September 2000, whereas the red line shows the average location of the sea-breeze front at 18:00 for the month of December 2000 (see color section)*

3.3.3. *Predicting rainfall in Mato Grosso*

The lack of a substantial network of rain gauges in Mato Grosso coupled with a series of average quality observations, means that satellite data must be used in order to carry out small-scale spatialization regarding the prediction of rainfall. With this information in mind, the use of satellite data for predicting rainfall in tropical regions has continued to increase over the past 20 years. These experiments have led to the creation of several international research programs [ARK 93].

Techniques using thermal infrared data were the first of their kind to be used [DUB 04]. These techniques are generally based on the indirect relationship between the temperature at the top of clouds and the intensity of precipitation that comes from these clouds. Algorithms are developed on the basis that a cold cloud summit means that the cloud itself is very thick, and is the origin behind convection rain that falls on the Earth's surface. The aim of this section of the chapter is to show how the infrared data, which is produced by geostationary GOES satellites, can be used to compensate for the poor observation network that exists in Mato Grosso. The data are then used to create a regional scale precipitation map for the climatological year 2000-2001.

The method used for the purposes of our study is made up of two datasets:

– The first dataset records the maximum temperature observed for each pixel on a daily basis. By analyzing such data over a period of 10 days or over a monthly period, elements such as atmospheric variables and clouds are eliminated, meaning it becomes possible to focus solely on the temperature on the ground. The temperature on the ground varies according to the type and amount of vegetation cover there is, and on the amount of water that has been absorbed by the vegetation (in other words the rain that has fallen). Studying variations in maximum brightness temperature comes back to the idea of measuring the quantities of water that have accumulated at one measurement site [GUI 94]. This method has been used in Africa as part of the Satellite Rainfall Estimation program. This program showed that there is a significant link between maximum brightness temperature and precipitation.

– A second dataset was created by thresholding the same images at a minimum temperature of –40°C: this meant that clouds with cold summits (convection clouds that produce rain) could also be monitored on a monthly basis (Figure 3.7). This method is an adaptation of the pioneering work carried out by Arkin on the Goes Precipitation Index, however, according to the results obtained by Cadet and Guillot in 1991 the threshold temperature is lower (–40°C in comparison to –38°C) [CAD 91].

The images in Figure 3.7 are used to monitor the development of clouds with a cold summit. The images are also used to see whether there is a link between these types of clouds and the quantity of precipitation that falls in the Mato Grosso region. The monthly correlation coefficients (Table 3.2) calculated for each month show that there is a link between satellite data and the quantity of precipitation that can be found on the ground.

Apart from this relationship, the link between precipitation and satellite data are not always very strong. The correlations are at their strongest at the beginning and the end of the rainy season, they are also at their strongest when the value of the maximum brightness temperature is at its highest, and this occurs half-way through the rainy season. Precipitation has a stronger correlation with clouds that have a cold summit than with maximum surface temperatures. If two satellite parameters are used to calculate multiple regressions and to predict precipitation, the correlation coefficients are above average for 7 months of the year ($r > 0.61$, which is 50% of the dependent variance). For 2 months of the year, however, the coefficients are just average.

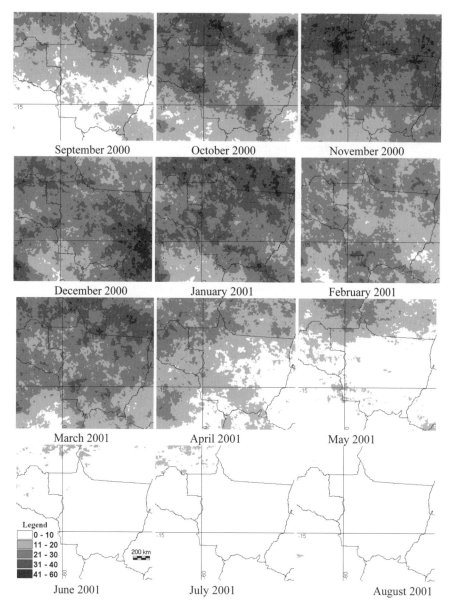

Figure 3.7. *The instances of clouds with a cold summit (GOES) in Mato Grosso: 2000-2001*

	Number of observations	Occurrence			Max Bright Temp			Occ+MBT
		R	A	b	R	a	b	R
2000 September	206	**0.61**	6.08	14.7	-0.35	-5.84	270.0	**0.61**
October	203	0.58	4.96	3.3	-0.44	-6.95	325.7	0.59
November	207	0.41	5.80	66.1	-0.40	-12.93	546.6	0.45
December	206	0.41	7.03	43.0	-0.48	-20.94	777.4	0.58
2001 January	201	0.52	12.68	-103.0	**-0.67**	-31.62	1032.7	**0.69**
February	200	0.31	6.49	64.3	**-0.66**	-29.78	944.3	**0.67**
March	198	0.46	7.37	65.1	-0.49	-23.09	798.6	0.53
April	196	**0.69**	9.47	-15.0	**-0.64**	-15.80	550.7	**0.71**
May	193	**0.66**	7.40	17.9	-0.39	-8.12	297.9	**0.66**
June	198	**0.66**	7.55	4.3	-0.57	-7.90	238.6	**0.67**
July	196	**0.69**	8.29	3.4	-0.48	-4.69	167.5	**0.69**
August	197	0.38	4.24	9.6	-0.06	-0.37	33.7	0.38
Season 2000-2001	2401	**0.78**	8.22	3.7	**-0.62**	-17.25	619.3	**0.81**
Annual Pmm	167	**0.81**	11.87	-659.7	-0.40	-60.01	3729.6	**0.82**

Obs. = number of stations used
Occ = monthly occurrences of clouds possessing cold summits
MaxBrightTemp = monthly maximum brightness temperature
Occ+MBT = multiple regression with occurrences and MaxBrightTemp
r = correlation; a and b = coefficients of linear straight lines.

Table 3.2. *Correlation coefficients (r) between rainfall measured on the ground and data produced by infrared GOES; coefficients with a value more than 0.60 are highlighted in bold*

Comparisons made between precipitation and infrared GOES satellite data in Mato Grosso for the period from September 2000 to August 2001 show that rainfall had a stronger correlation with clouds that had a cold summit (r = 0.78) than with maximum temperatures (r = -0.62). Precipitation estimation by satellite has therefore been slightly adapted in this region, a region that does not possess many conventional climatological stations. If we think of the two different methods that were introduced at the beginning of this section, the technique of precipitation estimation by satellite favors the use of the first method, as the second method only slightly increases multiple correlations (r = 0.82).

3.4. Examples of NOAA-AVHRR data used in Western France

3.4.1. *NOAA satellites*

NOAA orbiting satellites are American satellites. The NOAA program was created in 1970 and was the successor of the TIROS program. Today the NOAA program launches satellites into space approximately every 2 years. The NOAA satellites are located in low orbits in space, just 900 km above the surface of the Earth. They are referred to as orbiting satellites as they circle the Earth in a period of 100 minutes and their orbit is almost polar. There is a slight difference in the location of the ground track that is covered by the satellites and the actual meridians that exist on the Earth's surface. This means that it is possible for the satellites to fly over and monitor activity all over the planet, including high latitude areas, which is not the case for Meteosat satellites. In addition to this, the NOAA satellite is then capable of monitoring the same point at the same time every day. NOAA satellites are equipped with an AVHRR radiometer, which has a spatial resolution of approximately 1 km and is able to carry out measurements in six different channels that can be found in the satellite, with each channel responding to different needs [DSO 96]:

– 1: 0.58-0.68 µm, visible light (albedo);

– 2: 0.72-1.1 µm, near infrared (used to study vegetation);

– 3A: 1.58-1.64 µm, medium infrared (used to study ice and snow);

– 3B: 3.53-3.93 µm, medium infrared (used for the differentiation of clouds and snow/ice);

– 4: 10.3-11.3 µm, thermal infrared (used to study surface temperatures);

– 5: 11.5-12.5 µm, thermal infrared (used to study surface temperatures).

The longevity of this program, as well as the variety of applications created from recordings made by such applications, explain the importance of the scientific use of this data. These applications vary in size and can be used on a global scale, for example they can be used to monitor changes in vegetation cover brought about by human activity such as deforestation, or activities that influence climate. The applications can also be used on a regional scale and are also used on a daily basis for weather forecasting. All of this information will be shown through the examples of air temperature estimation and the monitoring of drought in the West of France.

3.4.2. *Estimating air temperatures in the region of Brittany, France*

Gathering information on local climates that are not equipped with a meteorological station means that values are recorded by a station in the vicinity and these are extrapolated and applied to the area that does not have a meteorological station. However, the values recorded at the nearby station need to be recorded in a similar environment. It is also possible to use a spatial interpolation method thanks to a grid of measurement sites created around the area that is being studied, in other words values from the surrounding areas are also recorded. Several different spatial interpolation methods are available and are commonly used. The first spatial interpolation methods used in France were made up of simple regressions between temperature and relief. These methods would then be improved by several different researchers, including those researchers who worked on the *Detailed Climate Map of France* [ER30 81]. Topographic factors, such as altitude, slope, and exposure are variables that continue to be used in these methods. In regions where there is not such a large contrast in relief, it is common to use interpolation methods that are less dependent on environmental factors, in other words the method of kriging is used. The work carried out in this field takes into consideration land use, which influences the energy exchanges that occur within the boundary layer. In this section a statistical method of interpolation used for maximum air temperatures in Brittany, France, will be presented. The images that were used for this particular method were NOAA-AVHRR images and they correspond to data that were gathered on April 10, 1997 [DUB 02]. The region of Brittany has interesting topography (even if the altitude level never exceeds 400 m), because of the existence of a particular peninsula. By studying this unique peninsula, the combined effects of latitude and distance from the coast can be easily explained. The diversity of land use makes it possible to use infrared data that are produced by NOAA-AVHRR satellites. For the purposes of our research these satellites had a spatial resolution of 1.1 km.

For this study 45 weather stations from the Météo-France network were chosen. For a region, which has a total surface area of 27,000 km^2, there was, on average, one measurement site for every 600 km^2. The study took place April 10, 1997 and occurred under anticyclone conditions. High pressure had established itself across North-Western Europe on this date, and all across France there were clear skies and a slight easterly wind, which brought masses of dry continental air to the region of Brittany. Anticyclone conditions are the best conditions for comparing and contrasting temperatures in relation to their environment. Slight winds mean that the air does not move and as a result this type of weather is likely to create local thermal contrasts.

The infrared radiation measurements carried out by the NOAA satellites are used to provide information about surface temperatures. The AVHRR radiometer, which can be found on board the NOAA-14, measures temperatures in wavelengths located within the ranges of 10.3 and 11.3 μm (channel 4), and between 11.5 and 12.5 μm (channel 5). It is, therefore, necessary to take into consideration the problems that are associated with atmospheric absorption and which disrupt the signal recorded by the satellite. This is why infrared radiation is used from channels 4 and 5 in the satellite. The split-window technique is a linear equation used to link the temperature of an unknown surface to a particular brightness temperature measured by the sensor. The use of this technique partially solves the problem that is associated with atmospheric absorption. There are many different techniques and formulae to be found in the literature [OTT 92], but for the purpose of this study the Vidal Madjar method has been used:

$$Ts = T4 + 2.78(T4\text{-}T5) - 1.35 \hspace{3cm} [3.1]$$

T4 and T5 are the brightness temperatures recorded in channels 4 and 5 of the AVHRR, and Ts is the temperature of the surface. The NOAA-AVHRR infrared data have to be considered as one of the variables that predicts air temperature. The two temperatures (air and surface temperature) should not be confused in any way. The other geographical variables used are latitude, longitude, distance from the coast, and altitude.

The results in Table 3.3 show that latitude is the variable that has a significant influence on daytime air temperatures. This is not particularly surprising as the study was carried out on a sunny day and the main variable that increases air temperature during the day is the amount of solar radiation emitted in relation to latitude. The fifth column of Table 3.3 shows the correlation after carrying out a multiple regression between the four above mentioned geographical variables as well as air temperature (Ta): the coefficient is very important and confirms that the temperature of the air does depend on latitude and distance from the coast. The second to last column shows that the correlation between NOAA surface temperatures and the Ta is identical to the correlation that exists for the four geographical variables and the Ta. This means that surface temperature recorded by satellites is a better indicator of air temperature than any of the four chosen geographical variables.

04-10-97	Altitude (alt)	Distance from the coast (dist)	Longitude (long)	Latitude (lat)	Four variables	Satellite (Ts)	Total
Correlation	0.26	0.58	0.36	-0.68	0.81	0.81	0.92

Table 3.3. *Correlations between air temperature (Ta) and geographical and satellite variables*

With the aim of developing an interpolation formula for air temperature, a multiple linear regression was carried out. It is therefore possible to express air temperature as the result of a linear combination with several variables. The formula which was developed for the 10th April 1997 is as follows:

$$Ta = 80.75 + 0.27Ts + 0.0028alt + 0.0303dist - 0.028lat - 0.0024long \quad [3.2]$$

Once the difference between the estimated temperatures (calculated by the model) and the actual temperatures that were measured on the 45 stations was calculated, differences of more than 2° accounted for only one case in 10 (9.6%). For two out of every three cases (between 62 and 71%) the estimation error that was generated by the model was less than 1°. The study of residues in relation to this model means that most of the 45 stations chosen as part of our research were located along the coast: the model seems to be better at producing spatial variations in temperature further inland, and this is due to the fact that there is a lot of mixed pixels near the coast. It is also possible to create an air temperature map from a multiple regression formula. A map, or more precisely a new image with a spatial resolution of 1 km is also created. This image shows that the air temperature value of each measurement site depends on the geographical characteristics of the site and on the surface temperature which is measured by the NOAA (Figure 3.8).

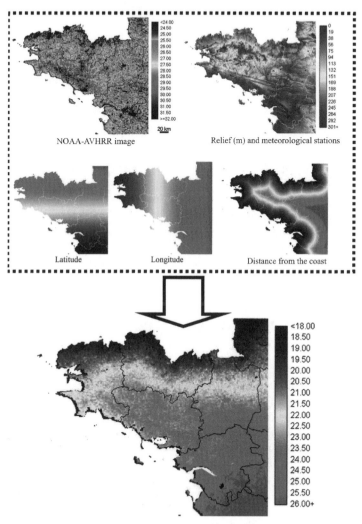

Figure 3.8. *An estimation of the maximum air temperatures taken on the April 10, 1997 in Brittany, France. The information was gathered from four geographical variables and NOAA-AVHRR data (see color section)*

3.4.3. *Monitoring drought in the region of Brittany by using the NDVI*

Summer droughts are a major constraint in the field of climatology. They affect water resources and cause problems as far as the growth of vegetation is concerned.

In agricultural regions drought can have serious repercussions on the region's economy. To monitor how vegetation cover is affected by drought it is possible to use data from NOAA satellites, from SPOT-Vegetation and from near-infrared images, in particular. The reflectance of the vegetation depends on the structure of the cells of the individual plants as well as on their water content. If a vegetation index (a combination of information produced by different spectral bands) is calculated for different dates of the year it becomes possible to compare vegetation cover for different times of the year. Amongst the most commonly used vegetation indexes used, the NDVI (normalized difference vegetation index) [ROU 74] is used for chlorophyllous vegetation (i.e. green plants) by using the red and near-infrared bands of the satellite's sensors:

$$NDVI = (Near\ infrared - Red) / (Near\ infrared + Red) \qquad [3.3]$$

In addition to cloud cover and the problems associated with atmospheric absorption [BAR 87], images have been synthesized by using a method known as the Maximum Value Composite (MVC) method. The most commonly used synthesis method is the method that synthesizes NDVI maximum values, and this usually takes place over a period of 10 days or over a monthly period. The MVC minimizes the presence of clouds and reduces the effects of sighting and solar angles. In addition, the MVC minimizes atmospheric effects caused by the presence of aerosols and water vapor [HOL 86]. At the moment MVC NDVI syntheses are commonly used to analyze vegetation cover on a global or continental scale. However, even though the MVC NDVI minimizes some effects that interfere with the radiometric values of the pixels, the effects are not completely removed from the images.

It is difficult to interpret the NOAA images due to the spatial resolution (1 km at nadir) of the AVHRR sensor. Monitoring vegetation cover using the NDVI is not always a straightforward process. The NDVI can sometimes be low due to a lack of vegetation (the vegetation may be stubble or ploughed) or due to the stage of the vegetation's life cycle. The NDVI can also be low because of the lack of water that the plants receive and this is reflected in their chlorophyll content. Therefore, it is necessary to have ground knowledge of the vegetation cover to be able to distinguish between the examples mentioned. This means that the NDVI cannot be considered as a drought map on its own, as a water deficits map could be. Conversely, the combined study of images taken during a year where there was no water deficit (calculated by the Turc's method and with an average of several years), compared with maps showing areas where there is a water deficit, shows that there are many different regions in France that experience abnormally dry periods. This is where images from low-resolution spatial satellites become important. These images make it possible to compare the spatial distribution of the regions by studying the vegetation cover of each region. This is made possible by using the water-deficit maps that have been created for each region.

During the summer of 2003, the high temperatures and lack of water limited vegetation growth. The NDVI NOAA AVHRR temporal series shows a strong decrease in the NDVI and a decrease in the total amount of vegetation cover that was present in agricultural areas from June 2003 onwards. This decrease, which only occurred in September in previous years, had never been as pronounced before (see Figure 3.9).

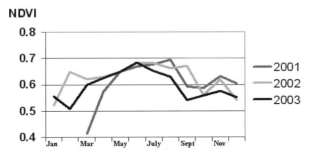

Figure 3.9. *A comparison of the annual NDVI NOAA-AVHRR values on surfaces mainly used to provide fodder for animals*

To carry out a more detailed study of droughts on a regional scale, it is necessary to associate the NDVI with the lack of water. The maps in Figure 3.10 show the water deficit (calculated by the Turc potential evapotranspiration formula and with a reserve of 125 mm) and monthly vegetation index (provided by NOAA-AVHRR data) for each of the 4 months in the summer of 2003. On the images it is possible to compare Western Brittany (with its low water deficit index and its high vegetation indexes) with Eastern Brittany, where the drought was more intense (especially in August) and where the vegetation indexes were much lower. The water deficit experienced in August (and which continued into the month of September in the East) was reflected in the NDVI values for September. This time delay of 1 month is due to the time taken for the vegetation to react to the stressful climate conditions and to the way in which the data were collected by monthly MVC.

The region of Brittany has the reputation of being a wet region. However, it can experience prolonged and pronounced dry periods, and in certain years, these dry periods can have a quite serious effect on the region's agricultural production. The detailed spatialization of a phenomenon as complex and diffuse as drought requires the use of data that has been produced by the process of remote sensing. The compatibility of the results produced by water analysis methods and by monitoring the NDVI leads us to believe that an inverse correlation exists between the NDVI value and the value corresponding to the water deficit. In theory, a high water deficit value corresponds to a low NDVI value, which would lead to the creation of drought. However, if the value of the water deficit is quite high then it becomes more difficult to analyze the NDVI value. It is necessary to take into consideration different the farming methods practiced, as well as the different types of vegetation, before carrying out any analysis on the variations of a specific satellite signal.

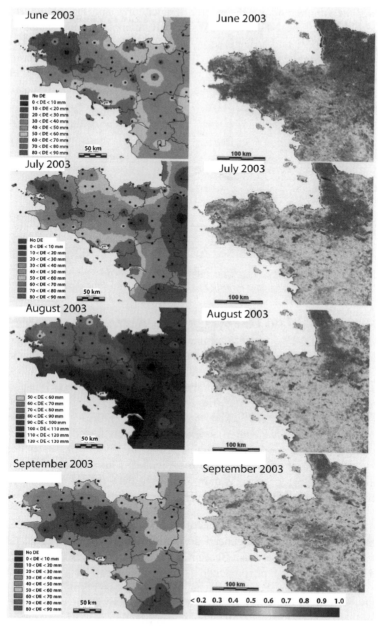

Figure 3.10. *Monitoring the water deficit (on the left) and the NDVI vegetation index (on the right) during the summer of 2003 in Brittany, France (see color section)*

Low-resolution NOAA-AVHRR NDVI data seem to be suitable for use on a regional scale. However, it is necessary to define the limits of any regional analysis as follows:

– spatial resolution, which is measured in kilometers, does not allows us to distinguish how drought affects one particular land use in comparison to another. The analysis is, therefore, carried out on a range of different crops. This means that it is not possible to specify which type of vegetation is affected the most by drought;

– a second point to be taken into consideration includes time delays, which occur as a the result of pre-processing methods used to produce the images. A monthly analysis of the NDVI will include index values recorded at the beginning of the month; yet during the entire period when the images were taken these values were steadily decreasing. There is a lack of precision during the analysis phase, especially from a spatial and temporal viewpoint. The use of medium resolution daily images (similar to MODIS type data) would limit the effects of such time delays and would limit the effects of using such a wide range of pixels.

3.5. Conclusion

The importance of using satellites in climatology and meteorology goes beyond the scope of research activities. Their usefulness no longer needs to be proved in numerous economic activities (agriculture, fishing, etc). The monitoring and forecasting of meteorological conditions (the cause of 20% of all road accidents in France) shape our daily lives, and the images produced by meteorological satellites are broadcast to the general public (it is impossible to have a televised weather update without the use of satellite animation). Satellite images produced by meteorological satellites are created every day [BAD 95; RAI 03]. Climatologists have a phenomenal amount of diverse information at their disposal. Our atmosphere is continually examined at many levels and there is a considerable amount of tools available to help climatologists carry out this task. A better understanding of the atmosphere will lead to models carrying out better weather forecasts in the future. A proper understanding of the different variables that affect climate makes it possible to slightly alter economic activity, to make slight alterations to buildings, etc. In addition, it informs people about what clothing they should wear on a particular day. In this respect, the climate is at the center of our everyday concerns, and strongly contributes to the characterization of a particular geographical area.

Satellite images have turned out to be fantastic tools used to spatialize geographical information that relates to climate. Homogenous spatial resolution and the repetitiveness of satellite data are precious tools that can be used to create detailed maps. New sensor programs should make it possible to increase spatial, spectral and temporal resolutions. The widespread use of GIS makes it possible to integrate satellite products as one of the entry layers in a GIS. Such an entry layer is necessary for providing spatial information about a particular piece of data. There are two important key elements to be remembered here: the satellite measurement is not equivalent to any measurement taken on the ground but merely provides extra and complementary information about it. Second, when carrying out any satellite measurements certain methodological precautions need to be adhered to (radiometric calibrations, geometric and atmospheric corrections). It should be noted that measuring surface conditions depends on cloud cover as far as visible light and infrared radiation are concerned. Nevertheless, satellite data are (and still will be in the future) essential support tools that are used to help spatialize climate data.

3.6. Acknowledgements

The author would like to thank the following people for all their hard work in the collaboration of this piece of work: Dominique Dagorne (Research Institute for Development, Lannion, France), Hervé Roquet and Jean-Pierre Olry (SATMOS, Lannion, France) for providing the satellite images; Anne Jallet, Aline Lecamus, Virginie Jumeau, Frédéric Damato, Olivier Planchon, Hervé Quenol, Rémi Lecerf and Pascal Gouery (COSTEL, Rennes 2); Josyane Ronchail (Research Institute for Development Brazil and Paris-7) as well as Andrelina de Santos and Waldemar Guimaraes (ANA-SIH) and Météo-France for providing the climatological information.

3.7. Bibliography

[ARK 93] ARKIN P.A., JANOWIAK J.E., "Tropical and subtropical precipitation", in R.J.Gurney, J.L.Foster, C.L.Parkinson, *Atlas of Satellite Observations Related to Global Change*, Cambridge University Press, p.165-180,1993.

[BAD 95] BADER J.M., FORBES G.S., GRANT J.R., LILLEY R.B.E., WATERS A.J., *Images in Weather Forecasting: a Practical Guide for Interpreting Satellite and Radar Imagery*, Cambridge University Press, 1995.

[BAR 87] BARIOU R., LECAMUS D., LE HENAFF F., *Corrections Atmosphériques,* Dossiers written on remote sensing n° 7, COSTEL - University of Rennes Press, 1987.

[BAR 74] BARRETT E.C., *Climatology from Satellites*, Methuen & Co, 1974.

[BUR 91] BURROUGHS W.J., *Watching the World's Weather*, Cambridge University Press, 1991.

[CAD 91] CADET D., GUILLOT B., *EPSA: Estimation des Pluies par Satellite*, ORSTOM, 1991.

[COR 06] CORPETTI T., PLANCHON O., DUBREUIL V., "Détection automatique du front de brise de mer sur des images satellites météorologiques", in *Proceedings from the XIXth conference held by the International Association for Climatology*, Epernay, p. 178-183, September, 2006.

[DSO 96] D'SOUZA G., BELWARD A.S., MALINGREAU J.P., *Advances in the Use of NOAA-AVHRR Data for Land Applications*, Kluwer Academic Publishers, 1996.

[DUB 94] DUBREUIL V., "La sécheresse dans la France de l'ouest: étude au moyen des bilans hydriques et des images des satellites NOAA-AVHRR", PhD thesis, University of Rennes-2, 2nd vol., 1994.

[DUB 02] DUBREUIL V., MONTGOBERT M., PLANCHON O., "Une méthode d'interpolation des températures de l'air en Bretagne: combinaison des paramètres géographiques et des mesures infrarouges NOAA-AVHRR", *Hommes et Terres du Nord*, 2002-1, pp. 26-39, 2002.

[DUB 04] DUBREUIL V., JALLET A., RONCHAIL J., MAITELLI G., "Estimation des précipitations par télédétection au Mato Grosso (Brésil)", *Ann. Int. Assoc. Climatol.*, vol. 1, pp.133-156, 2004 .

[ER30 81] ER. 30, sous la direction de PEGUY C.P, "Dix ans de carte climatique détaillée de la France au 1/250.000^e", in *Eaux et climats: Mélanges offerts à C.P. Peguy*, pp. 41-84, Grenoble, 1981.

[GUI 94] GUILLOT B., DAGORNE D., PENNARUN J., LAHUEC J.P., *Satellite et Surveillance du Climat, Atlas de Veille Climatique: 1986-1994*, ORSTOM Editions, 1994.

[GUR 93] GURNEY R.J., FOSTER J.L., PARKINSON C.L., *Atlas of Satellite Observations Related to Global Change*, Cambridge University Press, 1993.

[HOL 86] HOLBEN B. N., Characteristics of maximum-value composite images from temporal AVHRR data", *Int. J. Remote Sens.*, vol. 7, no. 11, pp. 1417-1434, 1986.

[KER 04] KERGOMARD C., "Réflexions sur trente années d'usage de la télédétection satellitale en climatologie", in *Proceedings from the XVIIth conference held by the International Association for Climatology*, Caen, p. 55-58, 2004.

[LER 83] LE ROY LADURIE E., *Histoire du Climat Depuis l'an Mil*, Coll. Champs, Flammarion, Paris, vol. 2, 1983.

[LOM 89] LOMBARDO, M.A., *A ilha de calor nas metrópole: O caso de São Paulo*. São Paulo, Hucitec, 1989.

[MAR 85] MARCHAND J.P., "Contraintes climatiques et espace géographique: le cas irlandais, Paradigme"; Caen, State thesis, 1985.

[MEN 95] MENDONÇA, F.A., "O clima e o planejamento urbano de cidades de porte médio e pequen : Proposição metodológica e sua aplicação à cidade de Londrina/PR", PhD thesis,University of São Paulo, 1995.

[MOU 81] MOUNIER J., LOZAC'H M., L'étude de la nébulosité au-dessus de la France à partir des photos de satellites", *Eaux et climats, Mélanges à C.P. Péguy*, Grenoble, p. 353-368, 1981.

[MOU 82] MOUNIER J., PAGNEY P., "Climats et satellites". *Ann. World Geogr*, vol. 505, pp. 275-299, 1982.

[MOU 90] MOUNIER J. (ed.), *Climatologie et télédétection*, International Association for Climatology, vol. 3, Rennes, 1990.

[OTT 92] OTTLÉ C., VIDAL-MADJAR D., "Estimation of land surface temperature with NOAA9 data", *Remote Sens. Environ.*, vol. 40, pp. 27-41, 1992.

[PAI 02] PAILLEUX J., "Les besoins en observations pour la prévision numérique du temps", *Météorol.*, vol. 39, pp. 29-35, 2002.

[PEG 89] PEGUY C. P., *Jeux et Enjeux du Climat*, Masson, Paris, 1989.

[PLA 06] PLANCHON O., DAMATO F., DUBREUIL V., GOUERY P., "A method of identifying and locating sea-breeze fronts in north-eastern Brazil by remote sensing", *Meteorol. Appl.*, vol. 113, pp.1-10, 2006.

[RAI 03] RAINER J.M., "Les satellites météorologiques", *Météorol.*, vol. 40, pp. 28-32, 2003.

[ROC 93] ROCHAS M., JAVELLE J.P., *La Météorol*, Syros, Paris, 1993.

[ROU 74] ROUSE J.W., HAAS R.H., DEERING D.W., SCHELL J.A., HARLAN J.C., *Monitoring the Vernal Advancement and Retrogradation (Green Wave Effect) of Natural Vegetation*, Greenbelt, Maryland, NASA/GSF, 1974.

[SCH 83] SCHIFFER R.A., ROSSOW W.B., "The International Satellite Cloud Climatology Program (ISCCP): the first project of the World Climate Research Program", *Bull. Am. Meteorol. Soc.*, vol. 64, pp. 779-784, 1983.

[TAB 89] TABEAUD M., "L'Atlantique tropical austral: l'eau atmosphérique et le climat en milieu océanique", PhD thesis, University of Paris IV, 1989.

[TON 76] TONNERRE-GUERIN M.A., "A propos de quelques techniques d'utilisation des images satellites en météorologie: contribution à l'analyse synoptique des latitudes moyennes de l'hémisphère nord", *Norois*, vol. 89, pp. 5-31, 1976.

[VER 92] VERGER F. (ed.), SOURBES I., GHIRARDI F., PALAZOT Y., *Atlas de Géographie de l'Espace*, Montpellier, SIDES-Reclus, 1992.

[WAH 97] WAHL L., "La dynamique spatio-temporelle des brouillards de rayonnement dans le Fossé Rhénan méridional à l'aide d'images NOAA-AVHRR", PhD thesis, University of Strasbourg, 1997.

Sources of satellite data

MSG Images: GDR CNRS MSG-ATR: (Second Generation Meteosat Research Group - Real-Time acquisition) PRODIG-University of Paris 7 and LMD http://prodig.univ-paris1.fr/msg/

GOES Images: SATMOS: Center for spatial meteorology, Météo-France, Lannion, France *www.satmos.meteo.fr*

NOAA Images: SAA: (NOAA Satellite Active Archive), *www.saa.noaa.gov*

TRMM Images: *Tropical Rainfall Mapping Mission*: http://disc.gsfc.nasa.gov/data/datapool/ TRMM_DP/01_Data_Products/02_Gridded/07_Monthly_Other_Data_Source_3B_43/

Chapter 4

Geographical Information for the Initialization of Numerical Weather Forecast Models and Climate Modeling

4.1. Introduction

Numerical models have been used operationally in weather prediction centers since the 1970s; since then, they have been subject to large improvements, either in terms of performances due to improved computer capabilities, increasingly precise descriptions of physical and chemical processes involved, improved horizontal and vertical resolutions and use of increasingly relevant observations [PAI 02].

Climate models which are derived from numerical weather prediction models generally include a much more detailed description of atmospheric composition. They have also improved dramatically, both for demonstrating climate predictability at monthly, seasonal or inter-annual timescales [PAL 04], and for the study of climate scenarios, i.e. the response of the climate system to different hypotheses on the evolution of the global socio-economical context. These different scenarios of our future climate have been the subject of assessment reports produced by the Intergovernmental Panel on Climate Change (IPCC). More information can be found in the 2007 fourth IPCC Assessment Report [IPC 07].

4.2. Brief description of the climate system

The climate system consists of five major components which interact at different time and space scales (see Figure 4.1):

– the atmosphere,

– the hydrosphere (liquid water at and underneath the Earth's surface: ocean, rivers, lakes, ground water),

Chapter written by Pierre BESSEMOULIN.

- the cryosphere (sea ice, glaciers, other frozen areas),
- the land surface,
- the biosphere (flora and fauna);

forced or influenced by various external mechanisms, the most important of which is radiation from the Sun.

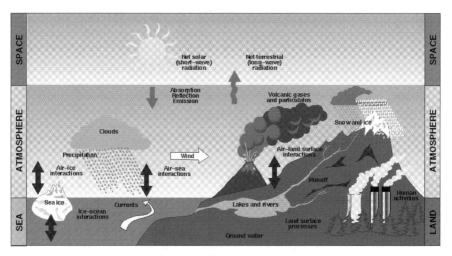

Figure 4.1. *The climate system*

The main drivers of the atmospheric and oceanic circulations are the rotation of the Earth, the atmospheric composition, the total energy received at the surface from the sun, and the radiative properties of the surface, and its topography and roughness.

The natural variability of the Earth climate is due to several variables:

– the variation with time of the Earth orbital parameters, which regulate e.g. the frequency of long-term ice-age periods (every 100,000 years) and which determine the characteristics of the interglacial periods. Such variables include the *obliquity* (measuring the tilt of the ecliptic compared to the celestial equator), the *eccentricity* of the Earth's orbit around the Sun, and the *climatic precession* which is related to the Earth-Sun distance at the summer solstice;

– the variations in solar activity, which modulate the amount of solar energy received at the Earth's surface, with typical timescales from years to decades;

– different oscillations, such as the ENSO (El Niño Southern Oscillation, and its two phases El Niño and La Niña), the NAO (North Atlantic Oscillation), etc., resulting from interactions between the ocean and the atmosphere, with timescales ranging from a few years to several decades;

– volcanic eruptions, which emit massive quantities of gases and aerosols into the atmosphere, and which are able to momentarily (for 1 to 2 years) change the amount of solar energy received at the Earth's surface, and thus affect temperature;

– it is also important to remember the annual and diurnal cycles, as well as the turbulent and chaotic nature of the Earth's atmosphere.

The role that the Earth's orbital parameters play (very well described by the Milankovitch theory) have been fully demonstrated with ice core data drilled into the 3,000m thick Antarctic ice sheets at Vostok station. It dated back 750,000 years and revealed the 7 past glacial cycles, and the close correlation between atmospheric greenhouse gases concentration (carbon dioxide, methane), and temperature.

According to Milankovitch theory, the expected future evolution of orbital parameters should have led a slow return to a small ice-age period six millennium later. This was consistent with the observed trend of global temperatures until the beginning of the industrial era, but is no longer the case, with an average increase over the 20th century of about 0.7°C, and current rate of approximately 0.2°C per decade, due to the impact of human activities.

Today, the total average amount of solar energy received as short wavelengths radiation by each square meter at the top of the Earth's atmosphere is 342 Watts per square meter (W/m^2), i.e. one quarter of the solar constant.

Figure 4.2 shows the Earth's radiation budget:
– 31% of the incoming energy received at the top of the atmosphere (107 W/m^2) is reflected back to space by the atmosphere (6%), clouds and aerosols (21%) and the Earth's surface (4%);
– the remaining 69% (235 W/m^2) are partially absorbed by the atmosphere and clouds (20% of these 69%, in other words 67 W/m^2), whilst the 49% complement (168 W/m^2) is absorbed by the Earth's surface.

The Earth's surface also emits energy as infrared radiation to the atmosphere as follows:
– 390 W/m^2, of which 40 W/m^2 are radiated directly to space;
– as sensible heat flux (7% of the incoming 342 W/m^2, i.e. 24 W/m^2);
– as latent heat flux (23% of the incoming 342 W/m^2, i.e. 78 W/m^2), corresponding to latent heat released when water vapor condenses in the atmosphere.

The radiative losses at the Earth's surface are compensated by a back radiation amount of 324 W/m^2 radiated by the atmosphere, especially due to water vapor and other greenhouse gases.

The condition for a stable climate is:
– a balance between the incoming solar energy (342 W/m^2), and the sum of the reflected solar radiation (107 W/m^2) and the outgoing to space long-wave radiation (235 W/m^2) emitted by the climate system;
– balanced short- and long-wave radiative budgets:

- in the atmosphere, radiative and convective losses are equal to 350-324-165-30+78+24=-67 W/m², which is balancing exactly the 67 W/m² fraction of solar energy absorbed by the atmosphere,

- at the surface, the solar contribution (168 W/m²) is counterbalanced by the sum of infrared radiative losses (390-324=66 W/m²) and by sensible and latent heat fluxes (24+78=102 W/m²).

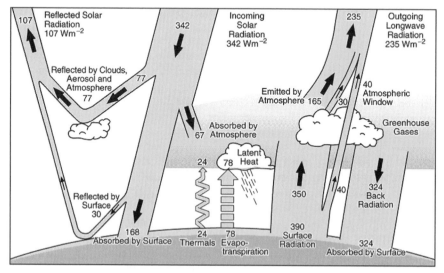

Figure. 4.2. *Average global radiation budget of the Earth's system. Source: "Climate change 2001 working group I: the scientific basis", IPCC Third Assessment Report*

This energy exchange occurring between the Earth's surface and the atmosphere maintains a global average temperature of 14°C on the Earth's surface. If there were no greenhouse gases present in the atmosphere, this average temperature would fall down to –19°C, a temperature corresponding to an infrared emission of the climate system, as seen from space, of 235 W/m².

The general circulation of the atmosphere and of the ocean, and consequently climate variables fields, are strongly constrained by the variability of the radiative fluxes (its diurnal and annual cycle, latitudinal variations, influence of cloud cover, GHG concentration).

Other relevant factors include altitude, type and properties of the underlying surface (especially radiative properties in terms of e.g. albedo), irrespective of whether the surface is land or ocean.

Geographical information, which provides boundary information used in weather and climate modeling is essential.

Meteorological phenomena cover a wide range of spatial and temporal scales, ranging from the micro-scale to planetary scale (Figure 4.3). They can be modeled explicitly when their scale is larger than the grid or resolution of the model. For those phenomena with a scale smaller than the model's grid, it is necessary to set physical parameterizations. The resolution of a model is determined by the size of the considered geographical area, the lead time of the prediction, the relevance of physical assumptions made and the available computer resources. This means that different types of models are used, depending on the objective. For example, the hydrostatic approximation is no longer valid on a small scale. Atmospheric processes that have non-hydrostatic effects include surface and atmospheric heat and moisture fluxes, turbulence, convection, evaporation, and condensation. For features less than 10-20 km, meteorological models must incorporate these non-hydrostatic processes.

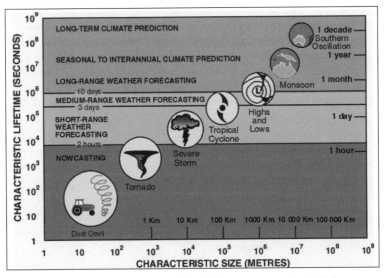

Figure 4.3. *Spatial and temporal scales of the main meteorological phenomena*

As an example, the resolution of operational weather forecast models is typically several tens of kilometers, whereas the resolution of a limited-area model is only a few kilometers, and the resolution of a global climate model, used e.g. to simulate the climate of one century, ranges from 200 to 300 km.

The geographical information that is to be used will therefore have to be adapted to these different resolutions. A value that is representative of a particular average on the model's grid as well as a value that describes the sub-grid variations will also have to be provided.

4.3. Brief overview of numerical weather forecast models

A numerical weather forecast model is a complex system comprising different phases and tools:

– *Pre-processing*: this phase includes carrying out quality control on the observed data that is exchanged globally through the Global Telecommunications System (GTS) implemented by the World Meteorological Organization (WMO). This phase also involves transforming the data that are to be used into the format required by the data assimilation scheme.

– *Objective analysis*: this is the phase enabling definition of an initial 3D representation of the climate system over the considered area (global or limited area), from data that are collected, and also from a recent forecast which is used as a first guess.

– *Data assimilation*: consists of establishing a series of analyses consistent with observations. The amount of observations taken into account is linked to the so-called cut-off time, i.e. the time allowed after the synoptic hour considered for collecting subsequent data. Data assimilation is meant to collect as much data as possible needed for describing, as well as possible, the initial state of the atmosphere and of other considered interacting components of the climate system, from which the forecast will be made. Most commonly used methods include optimal interpolation and variational assimilation. A special version of the latter has been developed to deal with non-synchronous observations (in other words 4D-VAR).

– *Physiographical data*: orography, land and sea areas, aerodynamic roughness, type of land and vegetation cover; surface albedo; etc, are geographical data that are generally maintained constant from the initial state to the final lead time for weather prediction purposes.

– *The initialization* is used to make the structure of observed meteorological fields consistent with the models that are used. It is also used to filter the development of oscillations generated by gravity waves, while retaining the most important meteorological structures. Commonly used techniques include digital filtering and normal-mode initialization.

– *The weather forecast model itself*: the evolution of atmospheric fluid is submitted to external forces (gravity, Coriolis force due to the Earth's rotation, and friction). It is also affected by internal forces, such as pressure differences and buoyancy. Both set of forces obey the laws of physics, as well as the laws of fluid dynamics, including conservation of mass and energy, and momentum.

In theory, the laws of temporal evolution of the different variables describing the structure of the atmospheric circulation make it possible to predict how the fluid will evolve, provided its initial state is known.

Due to uncertainties at all levels, ensembles of forecast are often used instead of a single forecast. For example, as the development of severe weather is frequently highly non-linear and therefore sensitive to forecast errors, this is an appropriate application of ensembles.

The simulations start with an initial state that is slightly perturbed in different ways, the produced ensemble is then subject to probabilistic analyses. The atmosphere is divided into a large number of elementary boxes (Figure 4.4) which cover either the entire world (global models), or a smaller zone (limited area model).

The dimensions of these boxes (which are equivalent to the model's grid) range currently in length from approximately several kilometers to several tens of kilometers. Their vertical dimensions range from several tens of meters in the lower layers of the atmosphere, up to several hundred meters in the higher levels.

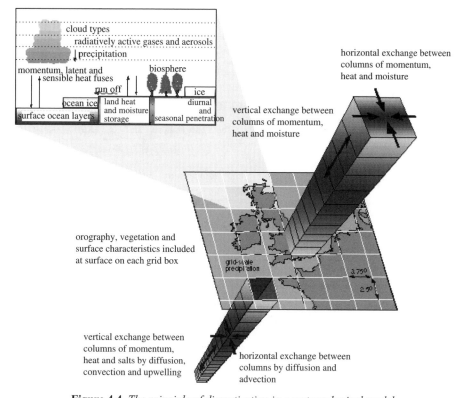

Figure 4.4. *The principle of discretization in a meteorological model*

The evolution of the meteorological variables (pressure, temperature, wind, etc) over time is integrated step by step at each grid point, starting from the initial conditions provided by the assimilation scheme.

Current numerical weather forecast lead times range from 6 hours (very short range weather forecasting) up to 10 days (medium range weather forecast), as done at the European Centre for Medium-range Weather Forecasts (ECMWF).

A model describes a certain number of processes that take place, including convection, cloud formation, precipitation, surface processes, radiation and turbulent exchanges. The model explicitly resolves processes that take place at scales larger than the model's grid scale, while sub-grid scale phenomena are parameterized.

The characteristics of the atmospheric component of the spectral model in operational use at ECMWF in August 2006 are displayed in Table 4.1.

Numerical scheme	T$_L$799L91 (triangular truncation, resolving up to wave number 799 in spectral space) 91 levels between the surface and 80km Formulation of integration time-step: semi-Lagrangian, semi-implicit
Time step	12 minutes
Equivalent grid resolution	Approximately 25 km
Minimal resolved wavelength	50 km
Number of grid points	76,757,590 in the atmosphere and 3,373,960 at and underneath the surface
Variables at each grid point	Wind components; temperature; humidity; fraction of cloud cover; water/ice content; surface pressure; ozone content
Included in the model	– Orography (altitude and sub-grid variation) – Four levels at the surface and in the sub-surface (allowing a description of vegetation cover, gravitational drainage, water exchanges occurring due to capillarity, surface and sub-surface runoff) – Convective and stratiform precipitation – Carbon dioxide (fixed at 345 ppmv); Ozone; Aerosols – Land surface and sub-surface temperature; sea surface temperature (SST); soil moisture; snowfall; snow cover and melting; incoming short wave and outgoing long-wave radiation, sub-grid scale orographic pressure drag; gravity waves and blocking effects; evapotranspiration, sensible and latent heat fluxes

Table 4.1. *Characteristics of the ECMWF numerical weather forecast spectral model*

Figure 4.5. *The grid used for the ARPEGE stretched forecast model with high resolution over Western Europe*

The global weather forecast model in use at Meteo-France known as ARPEGE has a variable horizontal resolution (see Figure 4.5). This makes it possible to provide a more detailed description of predicted weather over certain parts of the world. In its operational version, it has a resolution of 20 km over Western Europe, while over the antipodes resolution decreases down to 200 km.

Limited-area models, which are used to address regional weather or climate, are generally used in conjunction with global models, to which they are coupled, and from which they take their boundary conditions. They allow a better description and understanding of meteorological phenomena on a smaller scale than in global models. Examples of such models include: ALADIN and AROME in use at Meteo-France, HIRLAM, DWD COSMO and NCAR MM5 models, etc.

4.4. Role and description of the Earth's surface

4.4.1. *Continental surfaces*

The role of the surface in both nature and in numerical weather forecast models is to exchange energy and momentum with the atmosphere. The exchanges take place as fluxes of sensible heat, latent heat (evaporation) and momentum.

The different components of surface budgets (e.g. exchange of energy between the soil, the biosphere and the atmosphere) are described by specific models based on balanced:

– surface energy budget: $Rn = H + LE + G$

– surface radiative budget: $Rn = (1\text{-alpha})Rg + Eps (Rat\text{-sigma } Ts^4)$

– surface water balance: $P = LE + R + D$

where

– Rn is net radiation; H is sensible heat flux; LE is latent heat flux; G is sensible heat flux into the ground;

– alpha is the albedo of the ground; Rg is the global radiation; Eps is the emissivity of the ground with respect to the long wave downward radiation; Rat is the long wave downward radiation; sigma is the Boltzman constant; Ts is the surface temperature;

– P is precipitation; R is surface runoff; and D is a drainage term.

SVAT (soil-vegetation-atmosphere) transfer schemes such as the ISBA scheme in use at Meteo-France [NOI 96] have been developed to describe all surface processes. Given a description of the atmospheric forcing, and data describing soil and vegetation properties, they enable us to calculate net radiation, sensible and latent heat fluxes, soil heat flux, soil moisture, runoff and drainage and surface temperature.

These models generally need the following parameters:

i) parameters characterizing the soil:

– fraction of sand, clay and limestone present in the soil,

– soil depth;

ii) parameters characterizing the vegetation:

– vegetation fraction (veg= 1 – fraction of bare soil),

– the leaf area index (LAI),

– the normalized difference vegetation index (NDVI),

– the minimal stomatal resistance (Rsmin),

– the roughness length (Z_0, which is measured and expressed as a function of vegetation cover);

iii) Parameters characterizing both soil and vegetation:

– albedo and emissivity (both expressed as a function of the LAI).

4.4.2. *Ocean surface*

The roughness length is expressed by the Charnock formula for Z0, showing a quadratic dependence on the friction velocity.

4.4.3. *Observations*

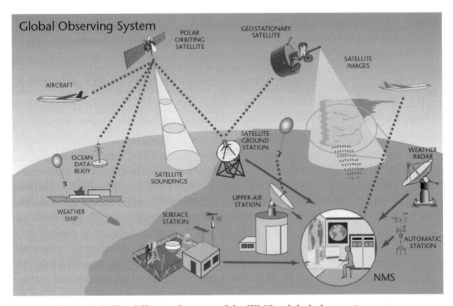

Figure 4.6. *The different elements of the WMOs global observation system*

Operational numerical prediction uses several types of conventional meteorological observations; however, nowadays, more and more observations are produced by satellite and are available in real time (see Figure 4.6).

As far as conventional observations are concerned, it is possible to distinguish the following three types of observations:

– surface observations are delivered by meteorological stations over land, by ships and buoys. The most useful measurements for numerical weather prediction are pressure, wind speed and direction at 10m, temperature and humidity at 2m above surface;

– radiosonde observations are generally carried out twice a day at precise times (00 and 12 UTC), and provide the models with information on vertical profiles of temperature, humidity and wind up to altitudes of approximately 30 km;

– airborne observations on board commercial airplanes, generally made by automatic systems, provide temperature and wind information along the flight route, and during take off and landing.

Amongst the satellite observations, it is possible to distinguish between the following types of observations:

– observations from atmospheric sounders aboard orbiting satellites: radiometric measurements (called radiances) provide information e.g. on the vertical profiles of temperature and humidity;

– wind observations using geostationary satellite (such as Meteosat) imagery are generally derived from monitoring the displacement of certain types of clouds.

The above mentioned observations are used by most of numerical forecast global models. As an example, the ECMWF is producing global analyses four times a day at 00, 06, 12, and 12 UTC, and 10-day forecasts from analyses made at 00 and 12 UTC.

The assimilation scheme performs the following operations:

Global analysis of:	Wind, temperature, surface pressure, humidity and ozone (4D variational assimilation called 4D-VAR which is carried out over 12 hour periods).
	Surface parameters: surface temperature from National Centers for Environmental Predictions (NCEP); sea ice derived from SSM/I satellite data; soil water content; snow depth and ocean waves.
Data used	In-situ conventional data over land and the ocean (SYNOP for the surface, TEMP and PILOT for upper-air profiles measured by radiosondes or rawinsondes, and tracked Pilot balloons; drifting buoys DRIBU data; AIREP, ACARS and AMDAR data from airliners and wind profilers)
	Satellite data (NOAA ATOVS eadiances; AIRS radiances; AMSU-A radiances; radiances from geostationary satellites and displacement vectors (derived e.g. from the displacement of clouds); surface wind from scatterometers; SSM/I radiances; ENVISAT total ozone column; SBUV ozone profiles)
	ENVISAT ASAR data for the oceanic wave model; JASON and ENVISAT altimetric data

Table 4.2. *Analyzed parameters and data used in the operational ECMWF assimilation scheme*

The following maps illustrate the data coverage of some types of these observations, for a typical synoptic time of observation (Figures 4.7 to 4.16). (http://www.ecmwf.int/products/forecasts/d/charts/monitoring/coverage/dcover/).

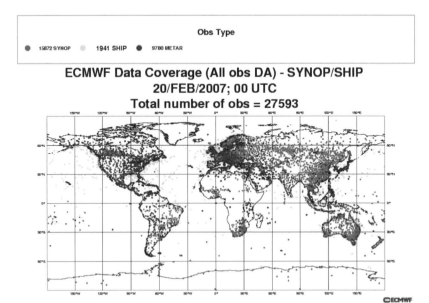

Figure 4.7. *SYNOP data used in the ECMWF analysis of 20/02/2007 at 00 UTC (see color section)*

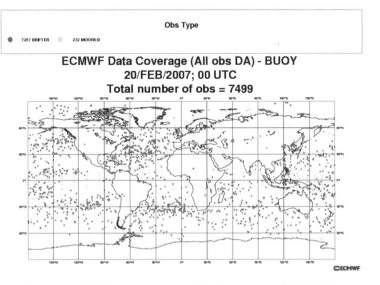

Figure 4.8. *BUOY data used in the ECMWF analysis of 20/02/2007 at 00 UTC (see color section)*

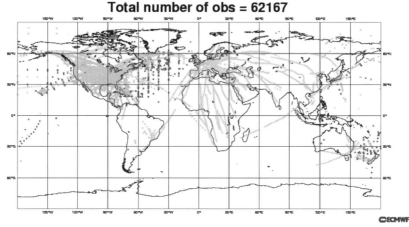

Figure 4.9. *AIREP, AMDAR and ACARS airborne data from commercial airplanes used in the ECMWF analysis of 20/02/2007 at 00 UTC (see color section)*

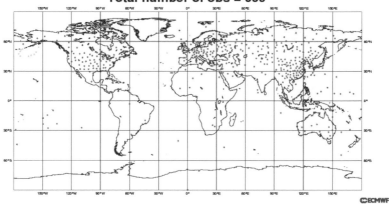

Figure 4.10. *TEMP radiosonde data used in ECMWF analysis of 20/02/2007 at 00 UTC (see color section)*

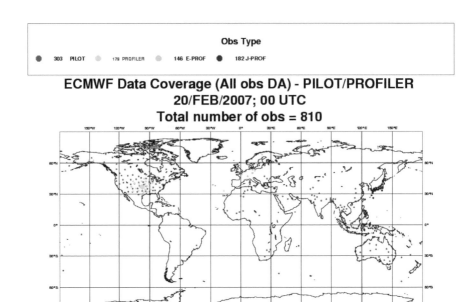

Figure 4.11. *PILOT or PROFILER vertical wind profile data used in the ECMWF analysis of 20/02/2007 at 00 UTC*

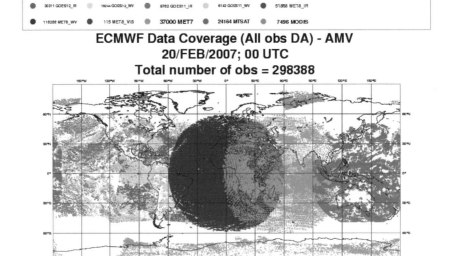

Figure 4.12. *SATOB geostationary satellite data used in ECMWF analysis of 20/02/2007 at 00 UTC*

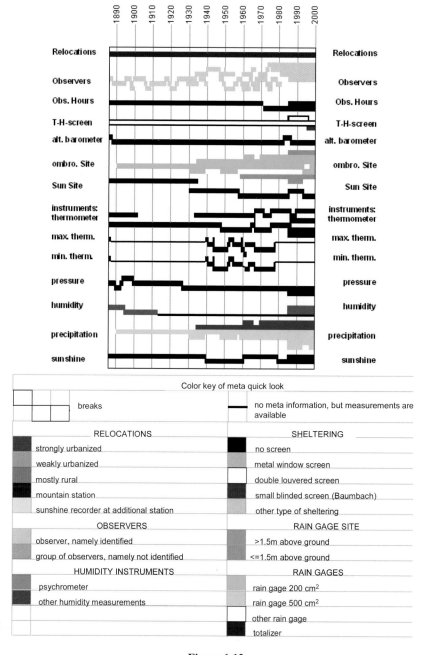

Figure 1.12

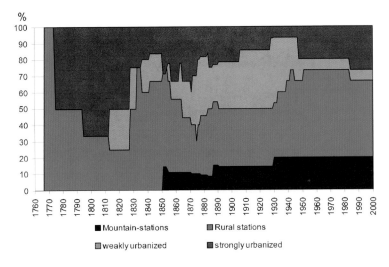

Figure 1.14

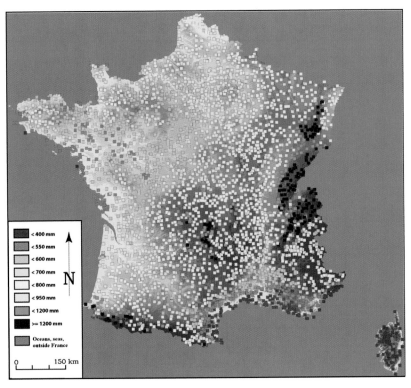

Figure 2.5

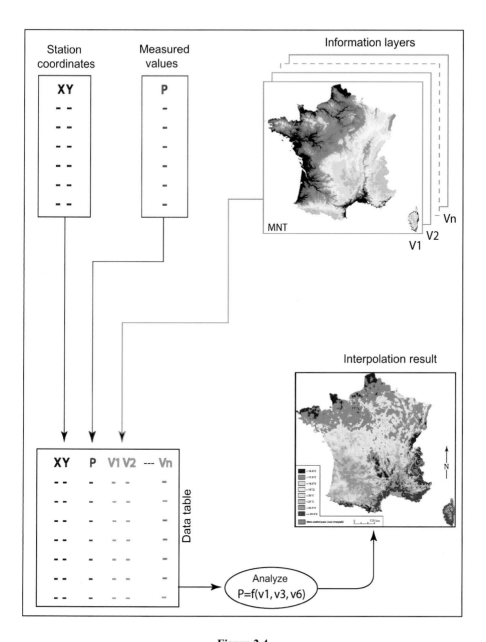

Station coordinates

Measured values

Information layers

XY

P

MNT

Vn

V2

V1

XY P V1 V2 --- Vn

Data table

Interpolation result

N

Analyze
P=f(v1, v3, v6)

Figure 2.4

Figure 3.5

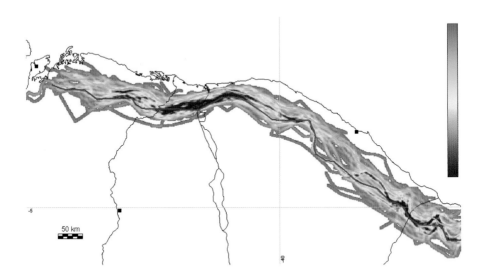

Figure 3.6

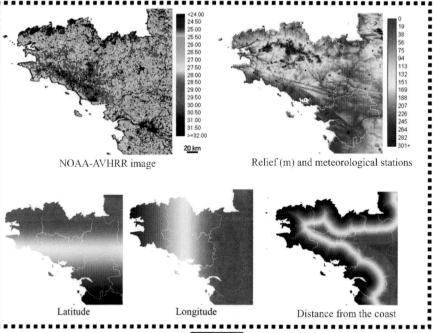

NOAA-AVHRR image   Relief (m) and meteorological stations

20 km

Latitude   Longitude   Distance from the coast

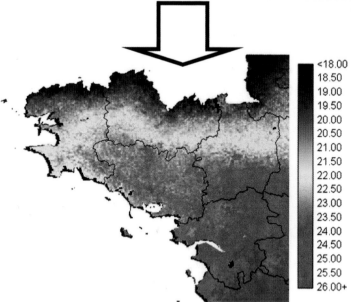

Figure 3.8

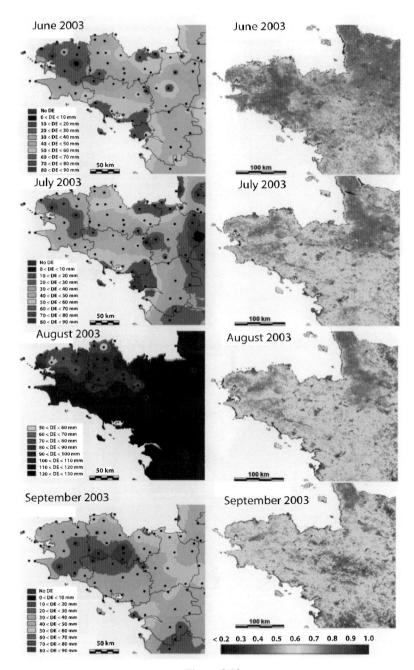

Figure 3.10

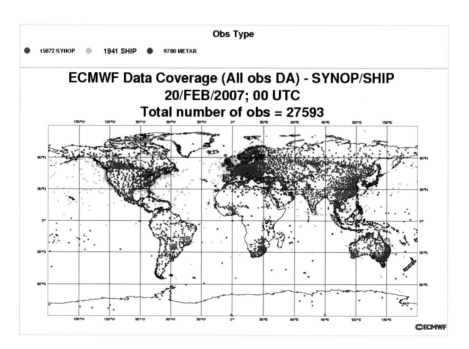

Figure 4.7

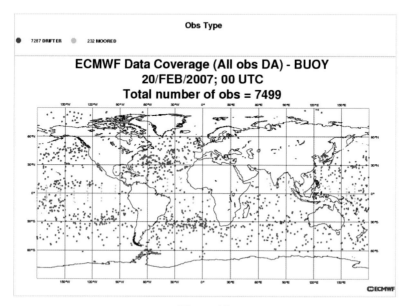

Figure 4.8

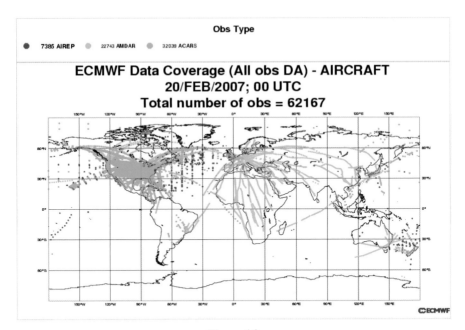

Figure 4.9

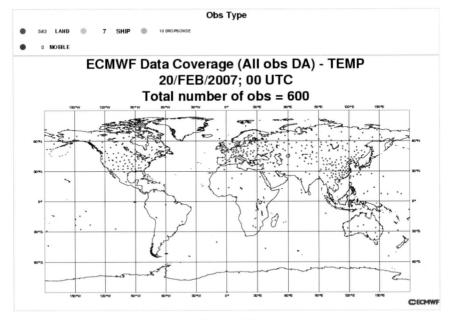

Figure 4.10

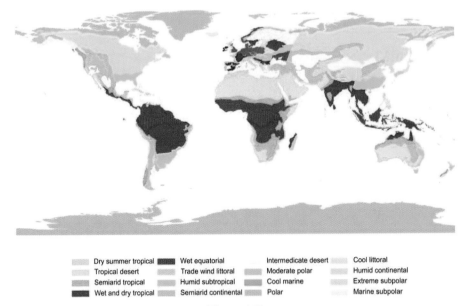

Dry summer tropical	Wet equatorial	Intermedicate desert	Cool littoral
Tropical desert	Trade wind littoral	Moderate polar	Humid continental
Semiarid tropical	Humid subtropical	Cool marine	Extreme subpolar
Wet and dry tropical	Semiarid continental	Polar	Marine subpolar

Figure 4.19

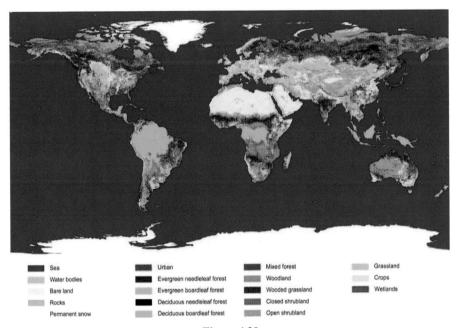

Sea	Urban	Mixed forest	Grassland
Water bodies	Evergreen needleleaf forest	Woodland	Crops
Bare land	Evergreen boardleaf forest	Wooded grassland	Wetlands
Rocks	Deciduous needleleaf forest	Closed shrubland	
Permanent snow	Deciduous boardleaf forest	Open shrubland	

Figure 4.20

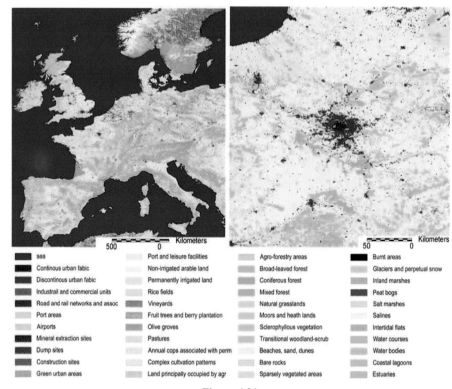

sea	Port and leisure facilities	Agro-forestry areas	Burnt areas
Continous urban fabic	Non-irrigated arable land	Broad-leaved forest	Glaciers and perpetual snow
Discontinous urban fabic	Permanently irrigated land	Coniferous forest	Inland marshes
Industrail and commercial units	Rice fields	Mixed forest	Peat bogs
Road and rail networks and assoc	Vineyards	Natural grasslands	Salt marshes
Port areas	Fruit trees and berry plantation	Moors and heath lands	Salines
Airports	Olive groves	Sclerophyllous vegetation	Intertidal flats
Mineral extraction sites	Pastures	Transitional woodland-scrub	Water courses
Dump sites	Annual cops associated with perm	Beaches, sand, dunes	Water bodies
Construction sites	Complex cultvation patterns	Bare rocks	Coastal lagoons
Green urban areas	Land principally occupied by agr	Sparsely vegetated areas	Estuaries

Figure 4.21

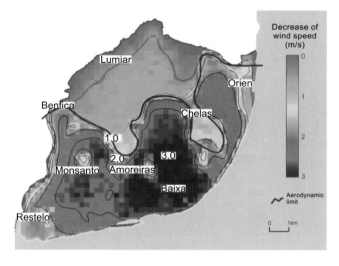

Figure 5.10

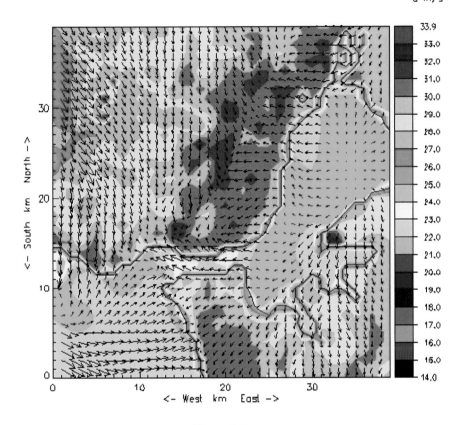

Tair at 2m and wind at 10m   2000/08/10 hour:12.0

6 m/s

33.9
33.0
32.0
31.0
30.0
29.0
28.0
27.0
26.0
25.0
24.0
23.0
22.0
21.0
20.0
19.0
18.0
17.0
16.0
15.0
14.0

<- South km North ->

<- West km East ->

Figure 5.11

a)   b)

Physiological
Equivalent
Temperature (°C)

12
14
16
18
20
22

Buildings

Figure 5.13

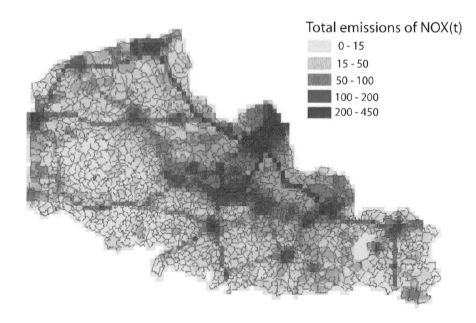

Figure 6.3

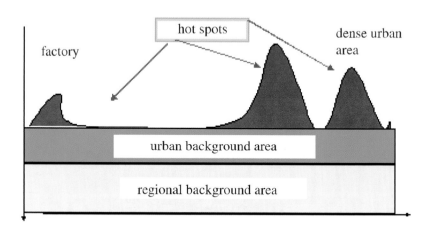

Figure 6.4

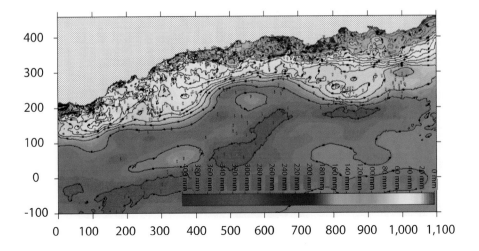

Figure 7.9

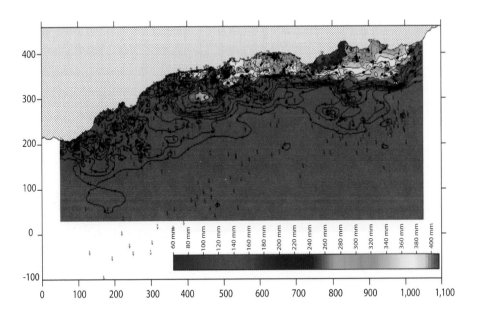

Figure 7.11

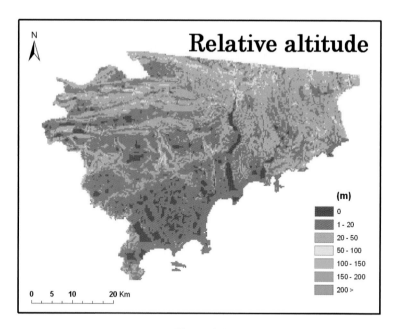

Figure 8.6

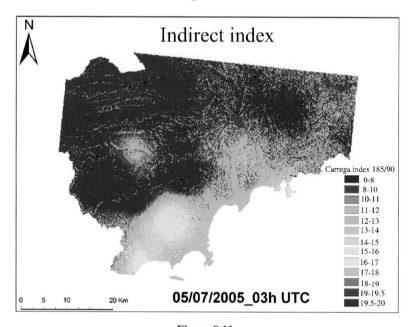

Figure 8.11

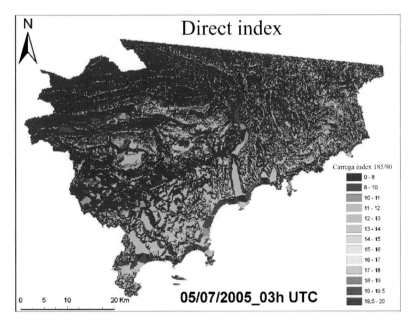

Figure 8.12

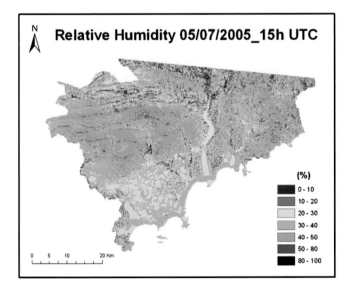

Figure 8.13

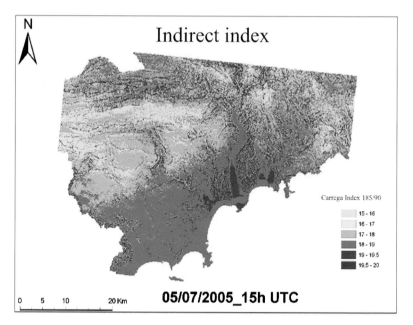

Figure 8.15

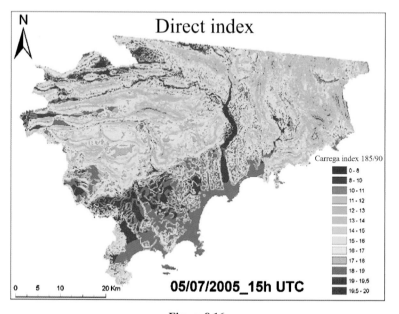

Figure 8.16

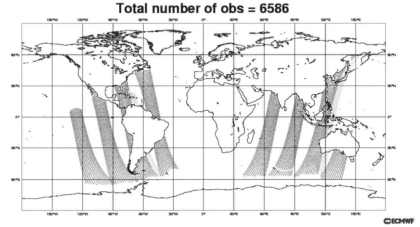

Figure 4.13. *SSM/I data from orbiting satellites used in the ECMWF analysis of 20/02/2007 at 00 UTC*

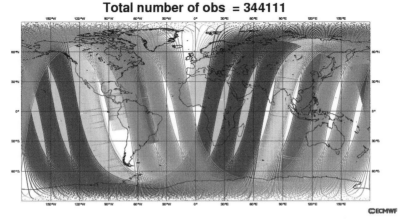

Figure 4.14. *ATOVS data from orbiting satellites used in the ECMWF analysis of 20/02/2007 at 00 UTC*

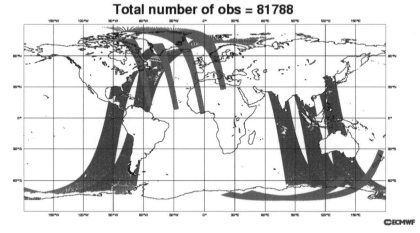

Figure 4.15. *Satellite SCAT scatterometer data (surface wind over ocean) used in the ECMWF analysis of 20/02/2007 at 00 UTC*

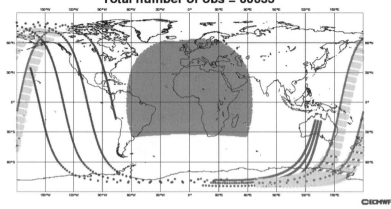

Figure 4.16. *Satellite ozone data used in the ECMWF analysis of 20/02/2007 at 00 UTC*

4.5. Description of surface parameters used in a forecast model

4.5.1. *Digital elevation models*

The majority of forecast models use the global elevation model GTOPO30, which was made available by the US Geological Survey Centre for Earth Resources Observation and Science (USGS/EROS) (http://edc.usgs.gov/products/elevation/gtopo30/gtopo30.html).

The data is divided into 27 geographical domains, which cover 50° of latitude and 40° of longitude, with a resolution of 30 seconds, which is equivalent to approximately 1 km (Figure 4.17).

Figure 4.17. *The GTOPO30 digital elevation model for Western Europe*

Different methods are used to compute the altitude of a model's grid:

– a grid-averaged orography;

– an "envelope orography" (one standard deviation of the subgrid-scale orography is added to the grid-averaged orography);

– a mixed approach, where averaged orography is kept at larger scales, and envelope orography is used at smaller scales.

Estimates of aerodynamic roughness length related to orography generally rely on a parameterization based on sub-grid variability of topography.

4.5.2. *Soil types data*

The FAO "Digital Soil Map of the World and Derived Soil Properties" (http://www.fao.org/icatalog/inter-e.htm) provides information on the different soil variables used in SVAT models. It does not cover all types of soils, but provides soil properties such as the surface albedo and textural data (fraction of sand, limestone and clay) at a resolution of 10 km, for three categories of soil color, texture and drainage [MAH 95].

4.5.3. *Land cover data*

4.5.3.1. *USGS data*

The USGS/EROS also provides information on vegetation cover at 1 km resolution, derived from AVHRR (Advanced Very High Resolution Radiometer on NOAA satellites). The data were first gathered between April 1992 and March 1993. It was first published in 1997 at global or continental scales, and then updated in 1998 (Figure 4.18) (http://edcsns17.cr.usgs.gov/glcc/).

The database gives access to:

– the different types of ecosystems classified either into a large number of classes, or into a limited number of classes consistent with those used in some SVAT models (e.g. 13 or 18 classes for the SIB model);

– the land cover classification (20 classes) adopted by the International Geosphere-Biosphere Programme (IGBP);

– seasonal variations of land cover.

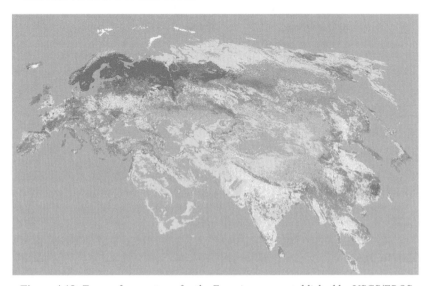

Figure 4.18. *Types of ecosystems for the Eurasian area established by USGS/EROS*

4.5.3.2. *The ECOCLIMAP database*

Météo-France [MAS 03] developed the ECOCLIMAP database with a resolution of 1 km (http://www.cnrm.meteo.fr/gmme/PROJETS/ECOCLIMAP/ page_ecoclimap.htm), with the aim of initializing SVAT models used in weather forecast or climate (global or regional) models.

215 ecosystems, which represent homogenous vegetation zones, have been derived from the combination of:

– existing climate maps (Koeppe and de Lond, 1958), giving 15 climate types at 1 km resolution (Figure 4.19);

– land cover maps: a global classification from University of Maryland into 15 classes at 1 km resolution (Figure 4.20);

– Corine Land Cover classification over Europe (Figure 4.21);

– and by using AVHRR data.

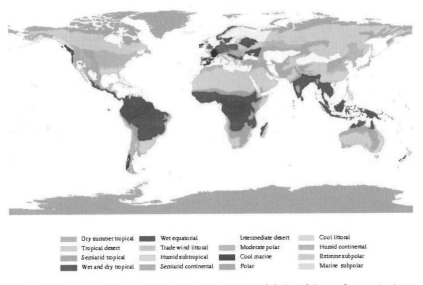

Dry summer tropical	Wet equatorial	Intermediate desert	Cool littoral
Tropical desert	Trade wind littoral	Moderate polar	Humid continental
Semiarid tropical	Humid subtropical	Cool marine	Extreme subpolar
Wet and dry tropical	Semiarid continental	Polar	Marine subpolar

Figure 4.19. *Global Climate Map from Koepp and de Lond (see color section)*

Surface variables used in SVAT models (such as LAI, vegetation fraction, roughness length, minimal stomatal resistance, albedo and emissivity) are computed using tables. These variables which are linked to land cover are derived every month from satellite data.

The very fine resolution achieved makes it possible to address regional and local scales. Such an approach is particularly helpful when nested models are used (in that case, the high resolution model takes its boundary conditions from the larger scale model). It ensures that the description of the surface in both models is made consistent.

Once sub-grid scale information is made available (e.g. fraction of bare soil) it is then possible to compute different energy budgets within the same grid. Different aggregation techniques have been developed and validated [NOI 95].

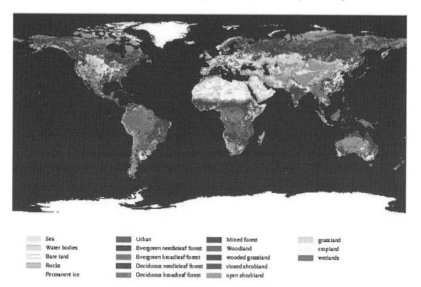

Sea	Urban	Mixed forest	grassland
Water bodies	Evergreen needleleaf forest	Woodland	cropland
Bare land	Evergreen broadleaf forest	wooded grassland	wetlands
Rocks	Deciduous needleleaf forest	closed shrubland	
Permanent ice	Deciduous broadleaf forest	open shrubland	

Figure 4.20. *Global land cover map from University of Maryland (see color section)*

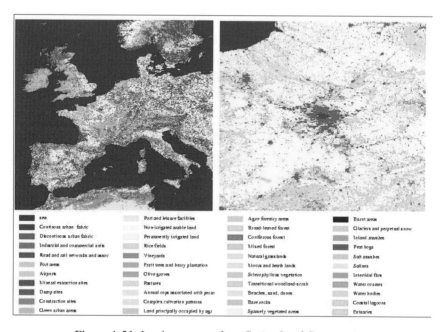

sea	Port and leisure facilities	Agro-forestry areas	Burnt areas
Continous urban fabric	Non-irrigated arable land	Broad-leaved forest	Glaciers and perpetual snow
Discontinous urban fabric	Permanently irrigated land	Coniferous forest	Inland marshes
Industrial and commercial units	Rice fields	Mixed forest	Peat bogs
Road and rail networks and assoc	Vineyards	Natural grasslands	Salt marshes
Port areas	Fruit trees and berry plantation	Moors and heath lands	Salines
Airports	Olive groves	Sclerophyllous vegetation	Intertidal flats
Mineral extraction sites	Pastures	Transitional woodland-scrub	Water courses
Dump sites	Annual crops associated with perm	Beaches, sand, dunes	Water bodies
Construction sites	Complex cultivation patterns	Bare rocks	Coastal lagoons
Green urban areas	Land principally occupied by agr	Sparsely vegetated areas	Estuaries

Figure 4. 21. *Land cover map from Corine Land Cover project over Europe (see color section)*

4.6. Bibliography

[IPC 07] IPCC: Fourth Assessment Report Climate Change 2007. Working Group I: The Scientific Basis, (available online at http://www.ipcc.ch/), 2007.

[KOE 58] KOEPPE C.-E., DE LONG G.C., *Weather and Climate*, New York, McGraw Hill, 1958.

[MAH 95] MAHFOUF J.-F., MANZI A.O., GIORDANI H., DEQUE M., "The land surface scheme ISBA within the Meteo-France Climate Model ARPEGE, Part 1: implementation and preliminary results", *J. Climate*, vol. 8, pp. 2039-2057, 1995.

[MAS 03] MASSON V., CHAMPEAUX J.L., CHAUVIN F., MERIGUET C., LACAZE R., "A global database of land surface parameters at 1-km resolution in meteorological and climate models", *J. Climate*, vol. 16, pp. 1261–1282, 2003.

[NOI 95] NOILHAN J., LACARRERE P., "GCM grid-scale evaporation from mesoscale modeling", *J. Climate*, vol. 8, pp. 206–223, 1995.

[NOI 96] NOILHAN J., MAHFOUF J.-F., "The ISBA land surface parameterization scheme", *Global and Plan. Change*, vol. 13, pp. 145-159, 1996.

[PAI 02] PAILLEUX J., "Les besoins en observation pour la prévision numérique du temps", *Météorologie*, 8th series, vol. 39, pp. 29-35, 2002.

[PAL 04] PALMER T.N. *et al.*, "Development of a European multi-model ensemble system for seasonal to inter-annual prediction", *Bull. Am. Meteorol, Soc.*, vol. 85, pp. 853–872, 2004.

.

Chapter 5

Assessing and Modeling the Urban Climate in Lisbon

5.1. Introduction

Cities cover less than 1.5% of the Earth's land surface. However, urban populations are currently growing at a much faster rate than the population as a whole, particularly in developing countries. Furthermore, mega-cities (with populations of 10 million or more) have brought about serious environmental problems [BRI 06]. The climate of a city depends on regional and local geographic characteristics, such as latitude, distance from water bodies, altitude, and topography, among others. Conversely, urban areas have a strong influence on natural environment, including climate, by modifying radiative, thermal, moisture and aerodynamic pre-existent characteristics. Consequently, there is a great variety of urban climates and a vast number of local and microclimates within each city, depending on air composition, urban morphology, and land cover (including the presence of green areas and water bodies). Urban climate casts an imprint on the surrounding areas, particularly downwind, but the influence of cities on "global" changes should be brought to attention as a considerable amount of greenhouse gases (GHG) are emitted in the urban environment [GRI 06a; ALC 08].

According to Oke [OKE 06a] there are several "modes" of research or practice employed by those involved in urban climate studies: 1) conceptualization; 2) theorization; 3) field observation; 4) modeling (statistical, scale, and numerical); 5) validation of models; 6) applications in urban design and planning; 7) impact assessment (post-implementation); 8) policy development and modification. Geographical Information Systems (GIS) are useful tools for different phases and levels of the research and application. This chapter will deal with the importance of geographical "support" in the correct monitoring of urban climate features (mode 3), in modeling parameters of urban climate at the meso and microscale and in validating some parameters (4 and 5). The objectives of these studies are to help implementing modes 6, 7, and 8.

Chapter written by Maria João ALCOFORADO.

5.2. Historical evolution of urban climate studies

5.2.1. *The beginning*

The influence of cities on air quality has been perceived since ancient times. Landsberg quotes Seneca's (ca. 3BC-65AD) remarking on the "heavy air of Rome" and the pestilent vapor and soot coming out of chimneys [LAN 81, p.3]. According to Yoshino [YOS 90/91] the Greeks and the Romans took into account the local climate conditions in the selection of the best sites for urban areas. Much later, London's frequent smog led to the prohibition by Queen Elizabeth I (1533-1603) of coal burning during Parliamentary sessions [LAN 81]. However, the influence of cities on climate is seldom referred to before the industrial revolution. The first quantification of cities effect on climate was made by the chemist Luke Howard in 1818 [LAN 81]. London was found to be 3°C warmer than the rural outskirts at night and 0.3°C cooler during the day. In 1855 and 1868, the Frenchman E. Renou pointed out the same night-time phenomenon in Paris [LAN 81]. He was aware that the differences were greater during clear, calm nights; that freezing nights were 40% more frequent in the rural areas; and that wind speed was lower in town. He also noted on the importance the correct location of instruments.

It should be reported that meteorological measurements began during the Renaissance a few decades after Galileo, Torricelli, and other scientists invented meteorological instruments, such as the thermometer and the barometer. Several early series (from the 17th century onwards) can be found in Italy [CAM 02], England [MAN 74], France [LEG 92; PFI 94], Germany [BRA 05] among others. However, it was a long time before the Academies of Sciences first and the World Meteorological Organization afterwards issued rules to standardize instruments and measuring conditions.

5.2.2. *The 20th century*

During the beginning of the 20th century monographic studies of urban climate in some cities were carried out. The effect of urban areas on climatic parameters was already known by the time the German priest Rudolf Kratzer wrote what we may consider the first manual on urban climatology (first edition in 1937) [KRA 37], in which he described several experiments that were being conducted by the scientific community. The influence of cities on thunderstorms was already known to Horton (1921, quoted by Shepherd) [SHE 05]. This research was continued by Atkinson [ATK 74] in London, Detwiller [DET 74] in Paris and in several US cities within the framework of Metromex (Metropolitan Meteorological Experiment) [CHA 74] during the 1970s.

The development of urban climatology was slow until the last decades of the 20th century, but a "boom" in the production of scientific experiments and papers has since taken place. This was partly due to fast development of technology (new sensors, computing power) and of new methodologies, as well as to the broad use of GIS. The last decades of the 20th century have also been very fertile in the theoretical frame of urban climatology. Synthesis from this period is included in the books by Landsberg [LAN 81] and Fezer [FEZ 95], in the chapter of Oke's book on the boundary layer climates [OKE 87], and in the papers by Oke [OKE 88] and Lowry [LOW 77; LOW 98], among many others.

5.2.3. *Urban climate in beginning of the 21st century*

During the first years of this century intense activity has been channeled into the development of monitoring techniques and modeling, into the study of urban climate in tropical and arid regions (or threatened by desertification) and in a wide dissemination of the latest results. Urban climate knowledge and methodologies have been broadly disseminated through the international conferences promoted by the International Association on Urban Climatology (IAUC) created in 2000 and its Newsletters (named *Urban Climate News* since March 2008) available at the IAUC internet site (http://www.urban-climate.org/) and through the International Urban Climate Homepage (http://www.stadtklima.de). Furthermore, some members of the IAUC have acted as guest editors for two special issues on urban climate of peer-reviewed international journals [*Theoretical and Applied Climatology* (vol. 83, 2006) and *International Journal of Climatology* (vol. 27, 2007)].

The present chapter is partly based on current methods, discussed in both journals and in recent literature, related to the assessment of urban climate. Modeling and the use of GIS are presented through examples involving Lisbon where a group of scholars have focused their research in urban climatology for the past two decades.

5.3. Spatial scales

The concept of scale is of paramount importance in urban climatology. Several horizontal and vertical spatial scales have been proposed by different authors. Oke's proposals [OKE 06a] will be followed in this Chapter, albeit in a simplified way.

Research on the urban environment may be carried out at different spatial scales, from the microscale to the mesoscale (Figure 5.1). The study of street, road, courtyard, and garden climate is carried out at the *microclimatic scale*, ranging typically from less than 1 m to hundreds of meters [OKE 06b]. The *local scale* translates the climate of city districts with similar built-up density, percentage of impervious surface, urban canyon geometry, activity, etc. One type of local climate (such as city-center climate) includes a mosaic of microclimates, and its signal may be captured in a meteorological station, integrating the microclimatic effects from nearby areas. Typical scales range from 1 to several kilometers [OKE 06b]. The mesoclimatic conditions of a town integrate several local climates. It can only be inferred from data provided by several stations scattered within the urban agglomeration. Typical scales are tens of kilometers [OKE 06b]. Examples of mesoclimatic and microclimatic approaches in the city of Lisbon are presented in 5.7 and 5.8.

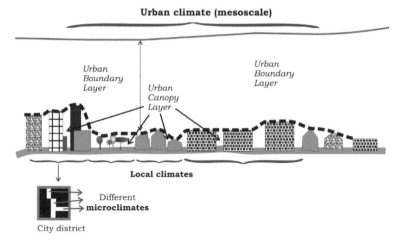

Figure 5.1. *Climatic scales and vertical layers in urban areas. Adapted from [AND 05]*

The influence of a city extends downwind into the suburban and rural areas as well as upwards. The most intense vertical exchanges of momentum, heat, and moisture occur in a layer comprised between the ground and the top of the buildings and trees. It is called the *Urban Canopy Layer* (UCL, Figure 5.1) and depends mainly of the microclimatic characteristics of the surrounding area. The *Urban Boundary Layer* (UBL) is the portion of the planetary boundary layer above the UCL; its climatic characteristics are modified by the urban area that lies below it [OKE 06b].

5.4. Climatic modifications induced by settlements

As the major modifications of climate by the urban environment have been presented in several papers and books previously referred to and updated in recent publications [MAT 01; ARN 03; KUT 04], only a brief overview will be given here.

Differences in the energy balance are responsible for air temperature spatial variation, particularly the well-known urban heat island (UHI). This term applies to the areas within the city in which the air or the surface temperatures are higher than those of the rural surroundings. The classical scheme by Oke [OKE 87] shows UCL air temperature rising from the suburbs towards the city center. The temperature increase is not regular but depends heavily on urban structure. The largest difference between urban and rural temperature is called UHI intensity. The UHI is more frequent and more intense during night-time in most cities and during the dry season in tropical urban areas [JAU 97]. In short, the main causes of the canopy layer UHI are the increased downwards long-wave radiation in town, decrease of long-wave radiation loss, high release of anthropogenic heat [SAI 04], less energy consumption in evapotranspiration, and the night-time release of heat stored during the day within the urban fabric [OKE 88].

The gradient wind is usually slowed down in the urban areas, although urban geometry may cause unwanted acceleration due to channeling along some streets or the *venturi* effect [CER 95]. Conversely, country and sea breezes may be intensified due to UHI. Cities are major sources of pollutants whose removal depends on the dispersion capability of the atmosphere [BAT 06] and on wind speed and direction. Figure 5.2a shows the streets of the city of Beijing (China) on a calm anticyclonic day of November 2005, while dense smog covered the city hampering the sun and causing health problems. The influence of the city on smog formation is still clearer if one compares Figure 5.2a with Figure 5.2b showing a view of the Great Wall taken the next day approximately 50 km from the urban area.

Figure 5.2. *a) Dense fog in the streets of the city of Beijing (China) on a calm anticyclonic day of November 2005. b) View of the Great Wall, taken the next day, approximately 50 km from the urban area*

Following the Metromex experiments and other studies quoted above it was verified that increases of 5-25% of rainfall took place downwind of the cities (in relation to the precipitation upwind) and that the rainfall anomalies increased with the size of the urban area [CHA 74]. The possible mechanisms of the urban impact in precipitation are the enhanced convergence due to high surface roughness in the cities and to urban heating, as well as enhanced aerosols in the UBL, functioning as "cloud condensation nuclei" [SHE 05]; as a consequence urban generated convective clouds may occur or convective clouds may be enhanced due to urban effect. As pointed out by Bornstein and Lin [BOR 00] and Shepherd [SHE 05] "bifurcating or diverting of precipitation systems" may also occur, but "to date there is no conclusive answer as to what mechanism dominates urban induced precipitation process or what the relative role, if any, of each mechanism. Furthermore, how the urban environment modifies these processes is also poorly understood".

The deeper knowledge recently acquired on urban climate is due to a great extent to the enormous progress in monitoring techniques, which have contributed the development of modeling methods. The results obtained need to be transformed or adapted to the different end-users. An overview of the urban climate monitoring and assessment methods is given in the next section.

5.5. Urban climate monitoring methods

5.5.1. *Mobile surveys*

In order to acquire detailed data, the first studies on climatology of the UCL were based on mobile measurements carried out on foot, by bicycle, motor bicycle, and eventually by car. The observers stopped regularly during the traverses and noted down the readings of meteorological parameters, using different types of devices. The *Assmann* psychrometer (Figure 5.3a) was one of the most accurate instruments for field observations of air temperature and humidity, despite its weight and slow performance. Nowadays, values are recorded through the use of digital devices during the mobile surveys (Figure 5.3b). Some research institutes use vehicles equipped with meteorological instrumentation that record and plot data continuously during the traverses, allowing for a much greater number and variety of sampling points [MAT 01; KUT 04].

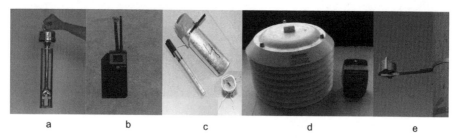

a   b   c   d   e

Figure 5.3. *Examples of instruments and shelters. a) Assmann psychrometer; b) portable digital thermo-hygrometer; c) temperature and humidity sensor Gemini data-logger and former shelter; d) plastic shelter and temperature data logger; e) plastic shelter located on a lamp at 3.5 m high*

Once sampling points are chosen, mobile surveys require decisions concerning survey's timing (frequency). Traditionally measurements were carried out three times a day: around noon (covering maximum temperature period), at dawn (frequently minimum temperature time) and some hours after sunset in order to sample the period when the intensity of UHI is at its highest.

5.5.2. *Fixed measurements*

Mobile surveys have two main disadvantages: their temporal and spatial discontinuity and the fact that the measurements cannot be carried out simultaneously in every point, so that corrections have to be introduced. Nowadays, "a set of data loggers installed in key points in towns has replaced most of the mobile surveys" [GRI 06b].

Some studies are based on pre-existing meteorological network recording continuously over time. However, meteorological stations (either traditional or automatic), following the rules of the World Meteorological Organization [WMO

96], provide useful data to understand the climate at the macroclimatic scale, rather than at the mesoclimatic or local scales. Furthermore, in most cases data are not accompanied by metadata (data about data). This means that often researchers only have access to long lists of figures and no information is provided as to exact location (latitude and longitude in degrees and minutes are insufficient), site characteristics and changes over time, instrumentation used, observation procedures and time of measurement. Moreover, as Grimmond [GRI 06b] pointed out stations from different institutions have different standards for instruments exposure, time of observation, and quality control of data. Furthermore, lack of information in urban stations about the instruments' location in relation to the street and building orientation, height of the instruments above the ground, percentage of impervious and green areas around the devices prevent firm conclusions [OKE 06b].

In some countries, the meteorological services have installed networks of urban meteorological stations. An "urban" station has been installed in Portugal for each 100,000 inhabitant of urban areas (http//www.meteo.pt). However, progress in the study of urban atmospheric processes has led to a greater awareness of the importance of correctly selecting the location of meteorological instrumentation in towns; often groups of scholars install their own networks in order to choose both location and device characteristics suitable for their research and expected results. Furthermore, improved instrumentation is now available to researchers due to the advances in technology, and it is possible to purchase reliable and accurate sensors and data loggers at affordable prices.

The "Initial guidance to obtain representative meteorological observation at urban sites" [OKE 06b, p. 2] has contributed to provide instructions for monitoring atmospheric parameters within the urban perimeter that are used in research at different scales, in order to understand the complex processes that occur in the urban atmosphere. As stated by T.R. Oke [OKE 06b] in urban areas "rigid rules have little utility", as "it is sometimes necessary to accept exposure over non-standard surfaces, at non-standard heights [....] or to be closer than usual to buildings or waste heat exhausts".

If the objective of establishing a meteorological network is clear, then reliable data can be obtained. Studies both at the mesoclimatic (settlement) and at the microclimatic scales have been carried out in Lisbon where two different networks were installed for the CLIMLIS project "Prescription of climatic principles for urban planning in Lisbon". Relying on Tim Oke's advice, consultant to this project, a mesoclimatic network (Figure 5.4b) was installed in order to detect air temperature variations due to the position of the measuring points in town, as independently as possible of the microclimate environment.

The first permanent data loggers, placed in meteorological shelters (Figure 5.3d-e), were installed in October 2004 following a first trial over shorter periods in which 12 measuring points were used. Seven are still active, recording data every 15 minutes and constitute the CEG (Centre for Geographical Studies of the University of Lisbon) "mesoscale measurement network". The immediate urban influence has been avoided by selecting measuring points with high sky-view factors (SVFs between 0.64 and 0.95) [ALC 06a]. The SVF is "the ratio of the amount of the sky 'seen' from a given point to that potentially available" (i.e. the proportion of the sky hemisphere subtended by a horizontal surface) [OKE 87, p. 404].

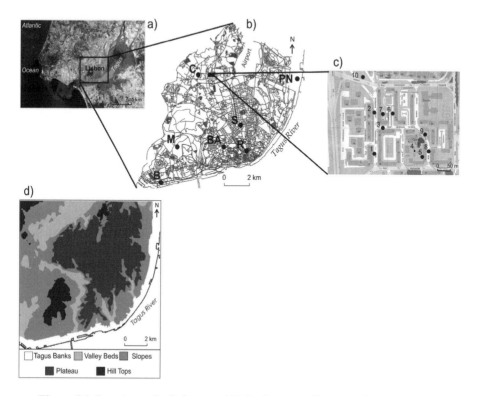

Figure 5.4. *Location and relief maps: a) Lisbon's region; b) mesoscale measurement network in Lisbon (Portugal), B=Belém, BA=Bairro Alto, C=Carnide, PN=Parque das Nações (Expo98), R=Restauradores, S=Saldanha; c) microscale network in Telheiras (Lisbon); d) relief map*

Examples are given in Figure 5.12 and will be discussed below. The measuring points are well distributed within the whole city and include sites on the outskirts (Monsanto Hill and Carnide northwards from Lisbon) and others directly on the Tagus bank (Parque das Nações to the east (E) and Belém in the southwestern part of the city) (Figure 5.4b). Measurements were taken at the height of 3.5 m, using Gemini Data Logger's Tiny Talk devices (Figure 5.3c-e). According to Oke [OKE 06b] and Nakamura [NAK 88] "measurements taken at heights of 3 to 5m differ little from those at the standard height, have slightly greater source areas and ensure that the sensor is beyond easy reach or damage" [OKE 06b, p.17].

In a second project designed to study spatial variation of climate features within a single city district, the data loggers were located in sites with different canyon geometry, street orientation, and land cover, so that the causes of microclimatic differences could be inferred (Figure 5.4c and 5.12) [AND 08].

5.5.3. *Remote sensing*

In this book, Chapter 3, written by V. Dubreuil, is dedicated to remote sensing. Additionally, there are a large number of publications on urban remote sensing [PAR 98; GRI 06b]. Satellite images have the ability of depicting several parameters of the urban areas with great detail (land cover, surface UHI, surface-atmosphere exchanges, radiation and energy balances, evaporation, carbon fluxes, among many others) and may be used combined with other techniques. Surface UHI and energy balance studies have been carried out in Lisbon [LOP 01; LOP 03] using remote sensing.

5.5.4. *Intensive measuring campaigns*

Large campaigns to monitor the urban climate are cited by Grimmond, who writes "increasingly, urban-based observational programs are collaborative, multi-institutional, multinational, interdisciplinary initiatives" [GRI 06b, p. 3]. These initiatives have the advantages of i) encouraging the research on processes and effects at different time and space scales; ii) being able to afford technically advanced instrumentation; iii) facilitating the common use of expensive instruments by several institutions; iv) conveying a greater awareness of the importance of appropriate placement of instrumentation; v) enabling faster and more sophisticated data analysis [GRI 06b].

Let us present data acquisition within the Escompte Project as an example. "The principal aim of the ESCOMPTE program was (..) to provide a comprehensive data base for the purpose of developing and validating models referring to air pollution in the region of Marseille, France, both at the mesoscale and the local scale". (http://www.cnrs.fr/cw/en/pres/compress/escompte.htm). However, in order to understand pollutants concentration, it is necessary to be acquainted with the UBL and suburban boundary layer dynamics. This has been carried out through an intensive measurement program using a) *ground-based instruments*, such as i) radars and sodars (Doppler acoustic sounders), ii) wind-profiling radars, iii) lidars in order to measure the quantity of transported ozone, iv) radiosounding, v) permanent chemical measuring stations, vi) many temporary stations (to measure climate, surface energy or chemical compounds), vi) a site equipped for the measurement of the principal meteorological parameters and chemical compounds (reference station at Le Planier); b) *air-borne instruments* such as equipped airplanes and constant altitude balloons launched in daytime, under coastal breeze conditions, to provide in-depth information on the extent of wind penetration into the mainland and into the areas affected by the pollutant combinations emanating from Marseille and the Fos-Berre industrial basin; c) *sea-borne instruments* aboard two ships equipped with chemical and atmospheric measurement tools (adapted from http://escompte.mediasfrance.org/projet/projets/CLU/).

5.6. Modeling

As there is a large spatial and temporal variability of urban climate parameters and phenomena, it is nearly always impossible to make use of a high-resolution network of meteorological stations and to monitor every single street, square, or park. Models have the ability to reproduce the spatial variation of meteorological features in urban areas, and sometimes even help to make predictions for the future. Furthermore, "their use opens new perspectives for example in the mitigation of UHI, or assessment of the role of air conditioning systems or the impact of urban dynamics on air pollution" [MAS 06]. Moreover, models are less costly and hence more adapted to the budget of most projects than intensive measuring campaigns and keys to interpretation of phenomena are supplied in several cases. However, models are not a substitute for measurements; data are necessary, not only to construct the models, but also to validate them. Several papers describe different types of models, give a number of examples, and refer to the potentialities and limitations of the different types of models [OOK 07; MAR 07; KAN 06; RAT 06]. Several models are described in http://www.stadtklima.de/EN/E_1tools.htm.

There are different types of models that can be applied to the urban climate: mathematical/physical numerical models, scale models, and empirical (statistical) models [HEL 99, simplified].

Mathematical numerical models are based on equations that reproduce the energetic processes that occur in urban areas and are able to simulate climate behavior in present and altered conditions. According to Pearlmutter [PEA 07, p.1877] they also "offer the flexibility to evaluate a wide range of urban configurations". However, the need of validation is not always considered by users of the different models. ENVImet (BRU 99) is an example of mathematical model that has been used in Lisbon's urban climate study. It is based on 3D "computer fluid dynamics" (CFD) and energy balance models, and it is widely used not only by urban climatologists, but also by architects. According to Martilli and Santiago [in BAK 08], CFD are "numerical models that solve the Navier-Stokes equation over small domains (few hundreds of meters at maximum), at high resolution (meters or less), and explicitly resolve the buildings". In the *Cost Report* edited by Baklanov *et al.* [BAK 08], several models applied to urban areas are described.

Scale models are considered valuable tools for characterizing the effects of detailed urban features and also for validating the predictions made by mathematical models. Kanda describes the major contribution of scale models (to study the statistical characteristics of turbulence and dispersion in neutral flows, the radiation balance, the nocturnal UHI, among others [KAN 06, p. 31]). However, according to Pearlmutter they cannot "replicate the complex interactions with other processes such as radiation and heat storage" or establish "the linkages between canopy-layer climate and the atmosphere above" [PEA 07]. Kanda also points out that this method "must be complemented with numerical models and field observations to overcome the mismatch between the thermal inertia of models and the real world" [KAN 06]. The wind tunnel model that we used is a scale model. It has helped us to understand

and visualize present flow within an urban district of Lisbon and to predict flow modifications, should the city continue to grow windward.

Empirical models are based on observation. The objective is to reproduce different parameters using statistical relations derived from observation [MAS 06]. "The approach is based on the assumption that the physical behavior is already contained on the observed data" [MAS 06]. Data from a great number of measuring points must be used. Several empirical models will be described for Lisbon, concerning the study of air temperature and "physiologically equivalent temperature" spatial variations at the mesoscale and at the microscale. Sometimes the procedure has been combined with other models (*ENVI-met*, *RayMan*). It is clear from the literature that a combination of models is used in most cases. "Using complementary approaches, there is reason to believe we can gain comprehensive understanding of turbulent flow, and the radiation and energy balances of the urban areas" [KAN 06].

5.7. Modeling Lisbon's urban climate at the mesoscale

5.7.1. *Empirical modeling and geographical information to estimate and interpolate near ground air temperature*

Detailed maps of different thermal patterns need to be constructed for various purposes, including the application of urban climate results [ALC 05; ALC 09] (http://pdm.cm-lisboa.pt/pdf/RPDMLisboa_avaliacao_climatica.pdf).

This has been carried out in Lisbon using a series of empirical models [AND 03] and adapting procedures previously followed by Carrega [CAR 92], Alcoforado [ALC 94], Joly and Fury [JOL 96], and Andrade [AND 98]. The factors influencing spatial variation of temperature were first detected by the way of a stepwise multiple regression [WIL 95]. Temperature data were then computed within a grid, and finally, thermal maps were produced [AND 03]. This methodology is similar to that described by Joly (see Chapter 2), but adapted to an urban area. The use of a GIS is indispensable for carrying out these procedures. Moreover, the GIS has the advantage to be easily updated [SVE 02].

The data used for this experiment were collected from 20 non-rainy days in September 2001 and in February 2002. Thirty minutes averaged temperatures were computed during night-time over the 20 investigated days. Air temperatures were standardized for each measuring point for each 30-minute period [ALC 06a]. As the parameters influencing air temperature vary over time, the first step involved building up groups consisting of days with thermal patterns. This was achieved using an automatic classification algorithm (Ward's minimum variance method) [WIL 95]. The algorithm identified seven air temperature patterns related to the same number of weather types. For each pattern, average temperature was first calculated, together with standardized temperature (T_{az}) for each measuring point.

In a second step, the standardized temperatures (T_{az}) were included in a multiple regression model as the dependent variable; the choice and the quantification of the independent variables were the result of several experiments [ALC 06a]. They have been computed within a 250 m grid around the measurement point. The use of mean values within each "unit of 250×250 m" leads to a certain loss of information but it also functions as a filter. Units of 100×100 m and 500×500 m were also tested and have lead to worse results. As the relations between T_{az} and the independent variables are often not linear, some of the latter had to be subjected to exponential, logarithmic, or other transformations [AND 03]. Multiple regressions of the standardized temperatures and different predictors were carried out separately for each thermal pattern. The coefficients of determination ranged between 0.68 and 0.92 [AND 03]. The most important predictors were: latitude, longitude, shorter distance to the Tagus, vegetation index (using normalized difference vegetation index (NDVI) factor, see Chapter 2), distance to the main axes, where tertiary activities are concentrated, percentage of built-up area within the grid corresponding to the measuring point and the product of the latter by mean building height of the same grid area.

In the third step, the spatial interpolation of air temperature for the whole city was carried out on a spatially-continuous base using a GIS [AND 98]. The values computed through the regression model were filtered (the value of each unit was replaced by the average of a group of 3×3 units (corresponding to a 750 m side square [AND 03, p. 177]). Finally, the isothermals were drawn based on the continuous thermal surfaces created. The three steps described have been repeated for each thermal pattern.

One of the patterns occurring during cloudless nights with moderate or strong north (N) and northwest (NW) wind (which is the prevailing wind in Lisbon) will be described. The independent variables that were found to be significant were: altitude, percentage of constructed areas, latitude, and distance to the Tagus (equation in [AND 03, p. 183]). The highest standardized temperatures (2.5°C) are projected to occur in densely built-up city districts located in bottom of valleys (Figure 5.5a). Highest temperatures extend to the N along the main streets. The lowest estimated temperatures occur in Monsanto probably due to altitude (253 m), vegetation, and absence of urban development. In this case, the maximum UHI intensity is 4°C. Topographic shelter is the main factor that influences the temperature distribution, and this is enhanced by the effects of urban density: altitude has the highest standardized regression coefficients (β), followed by the percentage of built-up area [AND 03]. In order to detect the urban effect, the model was run excluding the predictors related to urban properties. The resultant thermal pattern is shown in Figure 5.5b. In this case, the highest standardized temperatures predicted are lower: the core of the UHI, which was located over the high-density built up area and reached 2.5°C (Figure 5.5a), now attains only 1.5°C and has moved to the riverside areas of south (S) and southwest (SW) Lisbon (Figure 5.5b), where the topographic shelter effect is the highest. Thermal gradient in town is now only 3°C.

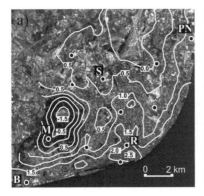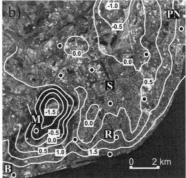

Figure 5.5. *Estimated standardized night-time temperature in Lisbon [AND 03]*
[ALC 06a]. a) In clear weather and moderate N wind; b) the same
but assuming the inexistence of the city (see text for details)

As shown, this procedure has not only enabled us to draw accurate detailed maps needed for different purposes, particularly in urban planning [ALC 05; ALC 09], but it has helped us to understand and to quantify the influence of different non-urban and urban parameters in near ground air temperature at the mesoclimatic scale. A particular downscaling based on further measuring and modeling will be presented in section 5.8.1.

5.7.2. *Frequency of thermal patterns. The Lisbon UHI*

The UHI is one of the most interesting features of urban climate and is considered a negative feature of urban climate in cities with hot summers and in the present scenarios of urban warming. However, UHI is not a static phenomenon. It is known to be more intense in calm and clear skies, but each city's thermal pattern has its own particular spatial and temporal dynamics.

This topic has also been studied in Lisbon, in order to derive results of use to city planners and stakeholders, by supplying information concerning the frequency of positive and negative features of the urban climate. The Lisbon mesoscale climatological network has supplied information continuously over time (since October 2004) and has helped to establish the frequency of the observed patterns. The definition of UHI as the difference between the highest temperature in town and the temperature of the rural area around it [OKE 87] may seem quite simple, but the methods to compute UHI have to be adapted to each single city and to the available data. In Lisbon's case it was calculated through the difference between the temperature recorded at the three measuring points located right at the city center (Saldanha, Bairro Alto, and Restauradores, Figure 5.4) and the two measuring points outside the urban perimeter (Monsanto and Carnide, Figure 5.4). Measuring points Belém and Parque das Nações were not considered once their riverside position confers them a particular type of climate.

The air UHI in the UCL (below the level of the buildings' roofs, Figure 5.1) occurs in 95% of the nights and 85% of the days, with median intensities between 1.7 and 2°C. It is more intense in summer and in situations with moderate W and NW winds, in opposition to the general rule of stronger UHI in windless situations. This is because Tagus and Ocean breezes frequently occur in Lisbon on windless situations cooling the southern districts. Daytime UHI in summer was less frequent at 2 pm than at 5 pm (Figure 5.6c) because at 2 pm the city is often affected by Tagus and Ocean breezes that reach Restauradores (R in Figures 5.3b and 5.5) and occasionally Saldanha (S, Figures 5.3b and 5.5); at 5 pm the breeze has often been replaced by the prevailing N or NW winds [ALC 87] (see section 5.7.3.2). The UHI intensity remains high all night long during spring and summer (Figure 5.6b), while in autumn and winter the UHI is at its highest intensity after sunset according to the general model [OKE 87].

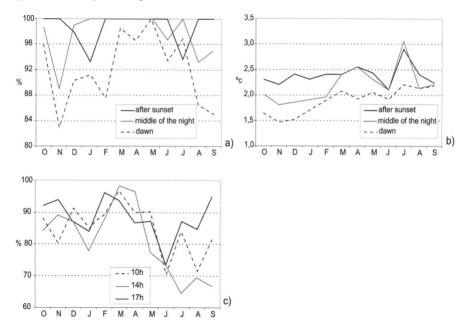

Figure 5.6. *UHI in Lisbon [ALC 07]. a) Frequency of nocturnal UHI; b) mean intensity of nocturnal UHI; c) frequency of diurnal UHI*

5.7.3. *Measuring and modeling wind over Lisbon*

5.7.3.1. *Prevailing N wind in the summer*

Wind is another weather parameter modified by urban areas. As shown in Figure 5.7, the prevailing winds at Lisbon Airport (located windward of the city, Figure 5.4b) are N and NW, particularly during the summer months. Urban roughness decreases air flux and, therefore, the urban logarithmic wind profile is different from that of an open area [MOR 93].

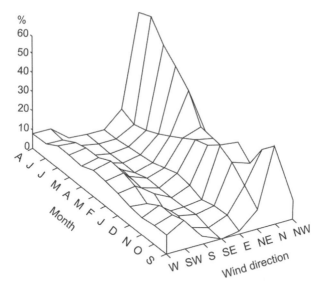

Figure 5.7. *Monthly frequency of wind
direction in Lisbon [ALC 92]*

In addition to a greater reduction of wind speed near the ground, the airflow is disturbed up to a higher altitude in an urban area that on rural landscape or over the ocean. The reduction of wind speed near the ground can be positive in cold climates or cold periods of the year in "temperate" climates, but it has several disadvantages for the city dweller. First, removal of air pollutants is less effective in calm weather (Figure 5.2a), and second, in warm periods, particularly during heat waves, the absence of wind aggravates the heat load of the individuals and can lead to discomfort; both situations may cause health problems. In most cities the knowledge of wind systems and patterns is, therefore, important for planners, decisions-makers, and the population [ALC 09]. Due to great temporal and spatial variability of wind speed and direction, to the fact that meteorological stations do not supply spatially continuous data, and owing to the complexity of urban areas, it is particularly difficult to understand spatial variation of wind patterns in cities. Numerical modeling can be a powerful tool to solve this problem. A study has been carried out in order to understand the summer winds in Lisbon.

3D meteorological models that can simulate and predict airflow and 3D temperature fields are suitable for meteorological simulation on particular conditions. Urban canopy parameterization was attempted in PSU/NCAR mesoscale model (MM5), but there are still several unsolved problems due to the complexity of the city geometry and heat fluxes within the urban fabric [RIG 08]. For the present purposes, a simple airflow model, such as WAsP, was considered appropriate. WAsP is a piece of software developed by Riso National Laboratory in Denmark, which estimates both profiles and spatial data taking into account the effect of terrain and surface roughness. It was shown that the model has good agreement with real wind climatology, confirmed by wind tunnel experiments [LOP 03].

Data consisted of wind speeds and direction averaged over 1 hour intervals collected at Lisbon Airport meteorological station (Figure 5.4b) for the period 1971 to 1980 [LOP 03]. WAsP transforms the original data into a set of statistical values that contains parameters of the Weibull distribution: A (scale, related to "mean wind speed") and k (shape) [PET 97], taking into account the topography and urban geometry. The accuracy of the estimations depends highly on the quality of data related to the aerodynamic roughness of the surface or "roughness length" (z_0). Z_0 is defined as "the height above the surface at which the mean wind speed becomes zero, when extrapolating the logarithmic wind-speed profile downwards through the surface layer" [OKE 87, p.57].

The same roughness length (z_0) values were assigned to areas that have similar urban morphology and built density. The procedure used to obtain the roughness map for the 1980s is explained in detail by Lopes [LOP 02] [LOP 03]. Z_0 highest values have been assigned to the central high density areas of the city ($z_0 = 1$) and to the hill of Monsanto ($z_0=0.7$) (Figure 5.8). Z_0 decreases although irregularly towards the NW, N, and NE of the city core, with low building density in the 1980s. The lowest value (0.01) was assigned to the airport area (Figure 5.8). Z_0 was computed for the 1980s, a period of rapid urban growth in Lisbon for which hourly wind data was available.

WAsP was run first *for all wind directions* in the summer (June to September). Wind profiles were estimated for 9 urban sites (representative of different urban geometry and position in town). In Figure 5.9, profiles 'a' and 'b' show average logarithmic wind profiles at different moments, whereas 'c', 'd' and 'e' show the profile referring to one single point: "Baixa" [near Restauradores (R) on Figure 5.4b]. The average profile 'a' corresponds to the period before the existence of the city, as the model was run considering only the topography, whereas profile 'b' refers to average wind profile estimated for the 1980s. In profile 'a', Weibull A parameter (that indicates wind speed) decreases from ca. 12 m/s at 250 m to ca. 6.2 m/s near the ground. However, building density in the 1980s caused a further decrease of 2 m/s near the ground (profile 'b').

When considering a single location: "Baixa", a still greater reduction in the 1980s is estimated of approximately 3 m/s wind speed (profile 'c'), when compared with profile 'a'. The use of WAsP provides the prediction of future wind-speed profile modification in central Lisbon (in "Baixa"), considering z_0 increase in the northern districts. A new dataset was prepared assuming that roughness will change from 0.01, 0.03, and 0.5 m in Northern Lisbon (at the time the original data were collected, Figure 5.8) to 1.5 m and to 2 m (typical z_0 of a city that increases in height and volume) [OKE 06b; MOR 93]. If z_0 raises to 1.5 m, then a further decline of wind speed will take place (profile 'd': minus ca. 4 m/s wind speed near the ground than in profile 'a'). A further increase in z_0 in the northern districts will not have a much greater impact (profile 'e'). Decreases of wind speed with increasing urbanization are expected at least up to 250 m high (Figure 5.9).

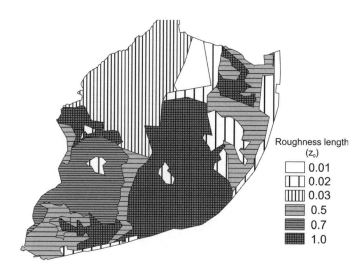

Figure 5.8. *Roughness length (z₀) in Lisbon computed for the 1980s, which served as input to the WAsP program [LOP 03; LOP SUB]*

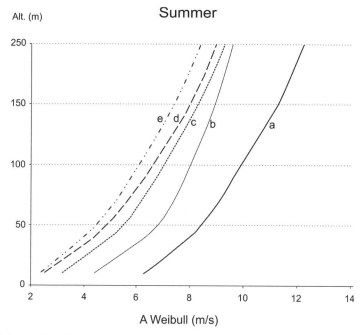

Figure 5.9. *Wind profiles in Lisbon for all wind directions simulated by WAsP [LOP 03]: a= simulation of wind profile over the site of Lisbon, excluding the roughness caused by urbanization; b=wind profile taking into account roughness caused by buildings in the 1980s; c=wind profile for "Baixa", southwards from Restauradores (R in Figure 5.4b) for the 1980s; d=projection of wind profile for Baixa considering z₀ increasing to 1.5 m in the northern part of Lisbon; e=the same considering z₀ increasing to 2 m in the northern part of Lisbon*

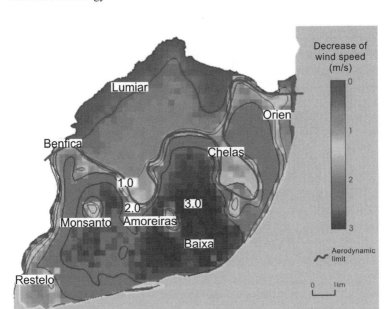

Figure 5.10. *Decrease of wind speed (m/s) at 10 m high during summer days of strong N wind (v>5 m/s) for the 1980s, simulated by WAsP software [LOP 03; LOP SUB] (see color section)*

As the N and NW winds are the prevailing wind directions, and there has been a very rapid growth of the northern city districts, wind speed will most probably tend to decrease in "Baixa" (near Restauradores, Figure 5.4b). In order to estimate the *summer moderate and strong* N wind modifications brought about by city growth, another wind dataset was obtained that contained only directions from 45° to 315° and speeds above 5 m/s (the WAsP software performed the necessary data modifications providing all the new parameters, i.e. frequencies, A, and k values) [LOP SUB]. By running this new dataset, Lopes represented spatially the average reduction of summer wind speed in the 1980s at several heights above the surface [LOP 03]. Figure 5.10 shows the reduction at 10 m high. The greater decrease occurs in "Baixa". The districts of eastern Lisbon, which were then less urbanized, suffer a smaller decrease. Wind speed is also reduced in the Monsanto hill because it is covered with forest and has therefore high z_0 values (Figure 5.9).

There are still some ventilation corridors in Lisbon that provide a renewal of clean and cooler air from the N [ALC 09]. Should these paths be obstructed a probable reduction of the air flux would occur and more frequent calm conditions would be observed. A subsequent impoverishment of air quality, as well as comfort and health for city dwellers, are to be expected.

5.7.3.2. *Summer wind breezes*

As mentioned before, the summer wind regime in Lisbon is dominated by the N and NW winds (Figure 5.6). Observational studies on summer winds in Lisbon and their influence on daytime regional and urban patterns have been presented by Alcoforado *et al.* [ALC 06b]. When the N and NW winds weaken by late morning or mid-afternoon, they are replaced by the Tagus and the Ocean breezes, which play a very important role in cooling the riverside city districts, as occurs in other seaside cities, such as Nice and Marseille [CAR 98; GAR 99]. During the heat wave of 2003, the temperature at Belém (B on Figure 5.4b) was 29.5°C, 11°C lower than at Restauradores in the inner city (R on Figure 5.4b) [ALC 06b]. It is known that breeze direction veers during the day [SIM 94]. In Lisbon it blows from the E and SE in the morning and veers to the S and SW in the afternoon [ALC 87; VAS 04a; ALC 06b].

For planning purposes, it is important to understand the timing and how the Tagus and Ocean breezes penetrate the city [ALC 09]. Owing to the difficulty in systematically measuring spatial wind-speed variations in urban areas, modeling was carried our by Prior *et al.* [PRI 01], Neto [NET 05] and Vasconcelos *et al.* [VAS 04a and b]. Prior *et al.* [PRI 01] used the *Topographic Vorticity Mesoscale Model* to study the boundary layer mesoscale flows on summer days in central and southern Portugal. He concluded that sea-breezes penetrate 80 km inland with wind speeds from 15-20 km/h and thickness between 400 and 600 m. Neto and Vasconcelos have used the high-resolution *Meso-NH atmospheric model*, based on an anelastic system of equations developed by Lipps and Hemler [LIP 82].

The details of the model dynamics may be found in Lafore *et al.* (1998), quoted by Neto [NET 05], who has adapted this methodology at the University of Évora (Portugal).The model simulates non-hydrostatic atmospheric movements and is able to simulate all scales ranging from turbulent large eddies to the synoptic scale. Through nesting procedures, it is possible to simulate local ventilation patterns in a 100×100 m grid [NET 05]. Maps of estimated wind direction in Lisbon at 6 am, at noon and at 5 pm during one warm summer day (August 10, 2000) have been presented by Vasvoncelos *et al.* [VAS 04b].

At dawn and during late afternoon N and NW winds are generalized within the whole area (not shown); however, breezes from the estuary reach a great part of the city at noon (Figure 5.11). The Tagus breeze is blowing from the E and SE over the eastern, southern, and central part of Lisbon, whereas in the western part of Lisbon's region, moderate wind from the NW is already blowing. Air temperatures estimated in the areas reached by the sea breeze are lower (under 30°C) than those in the western part of Lisbon (up to 33°C). Temperature measurements in summer have allowed us to validate this estimation. In fact, at noon fresh air reaches city districts a few kilometers away from the Tagus, as for example the University of Lisbon, southwards from the Airport, Figure 5.4b [ALC 06b; VAS 04a]. Despite having obtained interesting results, the authors are aware of the fact that geographical information at a convenient scale for the downscaling procedures is still missing [NET 05].

Toir at 2m and wind at 10m   2000/08/10 hour:12.0

$\longrightarrow$
6 m/s

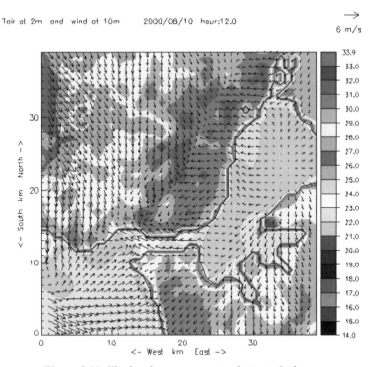

Figure 5.11. *Wind and temperature simulation in Lisbon on a summer day (August 10, 2000 at noon) [VAS 04a] (see color section)*

5.8. Modeling Lisbon's urban climate at the microscale (Telheiras city-district)

The microclimatic study was carried out at Telheiras (Figure 5.4c). The site was chosen because it is a flat 14 ha neighborhood of Northern Lisbon of approximately 100 m altitude and is located in an area where urban growth has been increasing since the end of the 20th century. Therefore, the results obtained here may be extrapolated to nearby areas. In this part of Telheiras there is a predominance of apartment buildings that form blocks in lines along the side of the streets; there are also some buildings taken up by social facilities and, to the W, a strip of "tower-like" buildings (Figure 5.4c). The main road axes are oriented in either N-S or the E-W directions and vary between 18 and 25 m in width. The height of the buildings also varies considerably: the tallest are around 25 m (32 m in the case of the towers). The H:W (height to width ratio) [LAN 81] varies between 0.5 and 1.1. The courtyards are taken up by parking spaces, social facilities and small garden areas.

The objective of this study was threefold: i) to study the microclimatic features and thermal comfort variation within the city district [AND 08]; ii) to verify modifications of wind speed and direction inside the city district [LOP 03]; iii) to prescribe climate guidelines for planning at the microscale. Special measurements, modeling, and GIS were used to develop the two first topics that will be presented here. The results are being used to prescribe climate guidelines for urban design and planning at the microscale.

5.8.1. *Spatial variation of physiologically equivalent temperature*

In order to prepare climate guidelines for urban planning the thermal comfort had to be assessed. The main objective was to make detailed maps and to assess how individuals would feel in the different parts of this neighborhood. The Physiologically Equivalent Temperature (PET) was chosen as an index for thermal bioclimate [HOP 99]. The discussion of the index and its application to Lisbon was performed in detail by Andrade [AND 03; AND 08]. PET has the advantage of being expressed in °C. It allows assessing the thermophysiologic combined influence of atmospheric parameters important for the energy balance of the human being [MAT 99]: air temperature (Ta), mean radiant temperature (MRT), vapor pressure (Pa), wind speed (v). The following parameters were considered constant: the production of internal heat (80 W/m^2) and the clothing insulation (0.9 Clo). For a set of actual atmospheric conditions, PET is equivalent to the air temperature in a room with a standard environment (Ta=MRT; v=0.1 m/s; Pa=12 hPa) that requires the same thermophysiologic response as an actual environment. For example, the combination of Ta=30°C, MRT=45°C, Pa=22 hPa, and v=2 m/s corresponds to a PET value of 34.1°C. This means that the actual thermal environment requires the same thermophysiologic response as a standard environment with Ta=34.1°C, MRT=34.1°C, v=0.1 m/s, and Pa=12 hPa.

The first step was the measurement of meteorological parameters performed by means of the "microclimatic network" (Figure 5.4c), devices and shelters similar to the ones used for the abovementioned mesoclimatic experiment (Figure 5.3d-e) and also some itinerary campaigns within the city district [AND 03; AND 08]. The sites where the sensors were placed ensured they represented the various micro-environments in the neighborhood.

The SVF of each site was calculated using the *RayMan* program [MAT 07] as an alternative to the use of fish-eye lenses. In the program, one can represent both the celestial hemisphere and horizon obstructions (trees and buildings) as well as sun trajectories at any specific date at a specific site. Figure 5.12 shows four different types of measuring sites in Telheiras. The calculation of SVFs by *RayMan* was verified by comparison with photographs with fish-eye lenses for 23 sites. The values of SVFs measured by both methods were not statistically different using analysis of variance (ANOVA; F=0.006, with a critical value of 4.05 for 0.95 probability) [AND 08]. There are clear differences between site 10 with a SVF of 0.74 in an open area and site 3 in a residential street with a SVF of 0.29. The solar paths at the solstices and equinoxes are shown in Figure 5.12. It is clear that site 3 stays in the shade during a great part of the year due to E-W orientation and low SVF. Site 2 in a N-S street with a 0.5 SVF, is in the sun during a large part of the day (with the exception of early morning and late afternoon) all the year round.

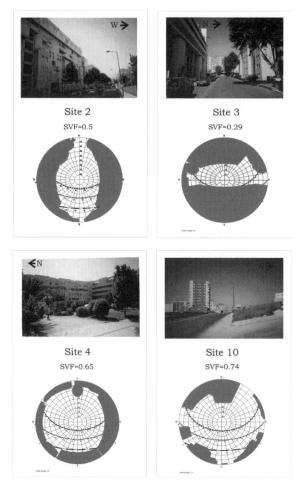

Figure 5.12. *Measurement sites 2, 3, 4 and 10 in Telheiras city-district,
SVFs, and solar diagrams, with sun paths at the time of the solstices
and the equinoxes [AND 08]*

In order to compute PET, wind speed and mean radiant temperature had to be previously modeled.

– Wind speed: as wind varies considerably both in time and in space, this parameter was assessed using the numeric model of microclimatic simulation called *ENVI-met* [BRU 98; BRU 99]. The simulation was carried out in a 3D grid that included buildings and vegetation. Wind speed and direction were estimated for each grid unit, for each value of speed and direction at the airport meteorological station (Lisboa/Gago Coutinho) [AND 08].

– MRT was modeled using *RayMan* [MAT 00; MAT 07] and the validation of this parameter was performed by comparing estimated and measured values [AND 08].

In short, PET was computed for different measuring points (1-10 on Figure 5.4c) using i) parameters measured in Telheiras during two 10-day measurement campaigns in winter and in the summer (temperature, humidity, short and long-wave radiation needed to compute MRT); and ii) parameters measured at the airport (solar radiation); iii) modeled parameters (MRT, wind speed). PET was modeled separately for daytime and night-time situations and in function of weather type [AND 03; AND 08]. Multiple regressions were calculated using PET as the independent variable and both geographical parameters and meteorological predictors at the airport as independent ones. One of the most important geographical parameters is the SVF. It is a very good descriptor of urban geometry and better than H:W ratio [LAN 81], particularly in irregular canyons, courtyards, and open spaces. Each independent variable was depicted in a different layer. Different equations were selected for night-time and daytime, and within the latter shade and sun situations, and winter and summer periods were separated. Independent variables were selected by stepwise multiple regression.

Two night-time situations will be described here. The best fit equation to night-time estimated PET values was the following:

$$\text{PET (t)} = 1.46184 + 1.07735\ Ta\ Airport - 3.07473\ SVF - 2.77031\ \ln V(t)+1,$$

where Ta is the air temperature and ln V(t) stands for natural logarithm of wind speed. The determination coefficient (r^2) attained 0.98, only 10% of the residuals are greater than 1°C and only 0.1% greater than 2°C. The β coefficients showed that the Ta measured at the airport is the main independent variable, followed by wind speed and by SVF. Therefore, one can consider that Ta at the airport represents the mesoscale thermal conditions to which variations in wind speed and SVF introduce microclimatic modifications. This model allows for the simulation of PET for any location in Telheiras and to other city districts in a similar geographic context.

Subsequently, spatial interpolation of PET among the measurement points was carried out using the regression equation determined in the previous stage. In this process a GIS with a 5×5 m pixel grid was used. Each independent variable corresponded to a separate layer. The estimated thermal pattern for two night-time summer conditions will be given as examples (Figure 5.13a-b). PET values between 18 and 23 were predominant and are considered comfortable according to Matzarakis *et al.* [MAT 99]. On a warm summer night, PET varies between 16 and more than 22°C (Figure 5.13a), i.e. most of the area is "comfortable". PET ranges from 12-18°C in a windy summer night (Figure 5.13b): conditions are "cool" (between 12 and 14°C) or "moderately cool" (between 14 and 16°C [MAT 99] [MAT 07]), and require an increase in the level of thermal insulation of clothing (close to 1.8 Clo) in order to maintain thermal balance of individuals. This indicates that even in summer, these areas are not always suitable for long stays outside without adequate protection when the N wind is blowing. The cooler areas lie in prevailing wind paths, which was confirmed by wind tunnel experiments (for example, the area marked (x) in Figure 5.13a and b, see next section and Figures 5.15 and 5.16). In both cases, the highest PET(t) values were found in areas sheltered from the wind and less exposed to the sky. The thermal behavior of the streets is found to be a function of wind speed. This is quite clear in simulation with the N wind, when the E-W oriented streets proved warmer than the N-S ones. In courtyards, PET is higher near the buildings, but lower in the middle of the courtyard. In the area of tall buildings to the W a mosaic of PET values are recorded, PET being higher in the "wind-shadow" parts of the buildings.

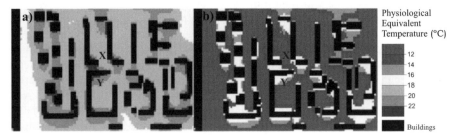

Figure 5.13. *PET estimated in Telheiras on two summer nights [AND 08]: a) warm night with light N wind; b) cooler night with moderate N wind (see color section)*

GIS modeling proved to be an adequate method to estimate the microclimatic conditions under different weather types, using only a small number of parameters and without requiring extra local measurements Assessment of the bioclimatic comfort in outdoor public spaces in Lisbon has been carried out in subsequent study [OLI 07] and in ongoing research in the frame of the Urbklim project (Climate and urban sustainability. Perception of comfort and climatic risks) (http://www.ceg.ul.pt/urbklim/index.html). In order to validate the use of PET in Portugal and to assess comfort perception, questionnaires were carried out simultaneously with weather measurement in several Lisbon green spaces.

5.8.2. *Wind tunnel experiments to detect main trajectories of prevailing winds*

Physical scale models permit the simulation of different meteorological parameters within controlled conditions. Kanda synthesized the advantages and the limitations of scale models [KAN 06].

A scale model of Telheiras (1:500) was built (Figure 5.14a and b) in the Portuguese National Laboratory of Civil Engineering (LNEC) in Lisbon for the study of wind speed and direction in this district of Lisbon. Several experiments were carried out during 1 month [LOP 03; LOP SUB]. The objectives were twofold: i) to study the airflow around the buildings and inside squares and inner courtyards; ii) to verify how the increasing roughness due to city growth in the windward side of the city would modify wind conditions in Telheiras district. In the aerodynamic wind tunnel an artificial air current is created (Figure 5.14c) in an open circuit aspiration tunnel with a 2 m × 3 m section, 10 m long. Wind speed inside the tunnel may be adjusted up to 18 m/s. The simulation of vorticity and roughness may be achieved through the use of different "separation elements" inserted in the tunnel (wooden blocks of different forms, dimension and space in between) windward from the scale model (Figure 5.14c). Several experiments were carried out for the three prevailing wind directions in Lisbon (N, NW, and W) and several combinations of blocks, simulating different roughness lengths (z_0, see above) were placed windward from the model and neutral atmospheric conditions were maintained inside the wind tunnel [LOP 03; LOP SUB]. Experiments were made for two z_0 windward from the model: $z_0 = 0.02$ m (current situation); $z_0 = 1.5$ m (if the city expands to the NW).

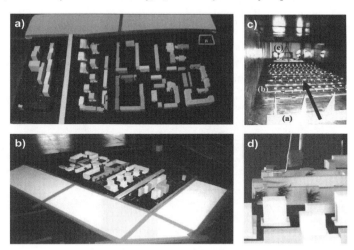

Figure 5.14. *Scale model and wind tunnel [LOP 03]: a) scale model seen from the S; b) scale model seen from the NW inside the wind tunnel; c) view of the interior of the wind tunnel, with wooden blocks that simulate the rugosity windwards of the model. They reproduce the conditions of the UBL. At the bottom the turbines and the scale model; d) detail of the scale model with the Pitot-Prandtl tube used to determine wind speed (another tube (not seen) at the top of the model takes measurements of the air layer not disturbed by the surface)*

Two examples have been selected from the different experiments carried out by Lopes [LOP 03]: indirect assessment of wind speed and direction through "erosion figure technique" and direct assessment of wind speed through the measurement of airflow inside the scale model. The objective of the "erosion figure technique" experiment is to monitor the influence of urban fabric on wind direction and speed. The ground of the scale model is covered with fine sand (diameter <6 µm) at the beginning of the experiment. The sand granules move through saltation. The observed "erosion patterns" that are visible on the scale model ground may decode relative wind speeds. Wind direction around and between the buildings was also assessed by observing, filming and photographing the scale model and the cotton threads placed inside it. Subsequently, wind speed is progressively increased inside to wind tunnel for different z_0 values upwind of the scale model, while measurements of airflow inside the model were carried out (Pitot-Prandtl tubes on Figure 5.14d), at different heights [LOP 03].

Expansion of the city

Two experiments were carried out for N wind situations in order to compare the current erosion capability of wind in the city district (with z_0 of 0.02 m upwind of the model) and a probable future situation within Telheiras, in which z_0 increases up to 1.5 m upwind of the model. Table 5.1 summarizes part of the results. The percentage of ground area of the scale model covered with sand at the end of the experiment confirms two facts:

– the percentage of sand staying on the ground of the scale model is higher when the simulation sets z_0=1.5 m upwind of the model than when the simulation sets z_0 at 0.02 m. This means that if the city continues to grow northwardly, then N wind circulation will be seriously affected in the city district, as the wind will not have the ability to remove the sand scattered on the floor of the scale model.

– G-shaped buildings are much more effective obstacles to N wind than tower-like buildings and L-shaped buildings.

Z_0 windward of Telheiras (m)	Tower-like buildings (N blocks)	L-shaped buildings	G-shaped buildings
Z_0 =0.02	0.5	4.4	14.7
Z_0= 1.5	3.2	16.8	37.2
Difference	+2.7	+12.4	+22.5

Table 5.1. *Percentage of ground area of the scale model covered with sand at the end of the simulation for N wind [LOP 03, p.212]*

Figure 5.15. *Erosion figures experiment inside a scale model representing Telheiras city district [LOP 03]. The figure refers to the moment when the wind direction had attained 9.3 m/s in an undisturbed area on the "ceiling" of the wind tunnel*

Figure 5.15 shows a photograph of sand distribution during an experiment considering a roughness length of 1.5 m windward of the city district. At the moment the photograph was taken, wind speed had increased to 9.3 m/s in an undisturbed area on the ceiling of the wind tunnel (Figure 5.14d) [LOP 03].

Different urban fabrics react differently to wind. When looking at the "tower-shaped" buildings in the western part of Telheiras, it is clear that ventilation is slightly hindered by this kind of building organization, and only the southernmost blocks originate some sheltered areas (where the sand has remained on the ground). I- and L-shaped buildings create wind corridors, such as x on Figure 5.16, where the

N wind is systematically channeled. If we look back to Figure 5.13 (e.g. point x), where PET is expressed, we see that these are the areas that remain relatively cool in warm nights and also cool on windy summer nights due to higher wind speed. For PET calculation, wind was modeled using another software (*ENVI-met*) and when wind is channeled its speed increases; so the wind tunnel experiment has also helped to validate *ENVI-met* simulations. G-shaped buildings create particularly sheltered areas. Inside the courtyards (such as y on Figure 5.16) airflow is trapped and there is no air renewal; there will only be eddies and recirculation of the same air, that may move pollutants and particulate matter. As some of these courtyards are used for car parking, one may guess that the air quality inside them will frequently be poor (no measurements to prove it as yet).

Numerous other experiments have been conducted by Lopes [LOP 03] for the three prevailing wind directions: N, NW, and W. Figure 5.16 synthesizes the main airflows and indicates clearly the areas where circulation is more disturbed. In most cases, airflow within the city district is similar when gradient wind blows from the N and NW, so the same arrows refer to both gradient wind directions. Western gradient wind originates other flow directions in the city district. It is noticed that the worst ventilated areas (inside G-shaped buildings) correspond to the areas where sand accumulation persisted during the experiments and where whirls as well as (bad quality) air recirculation occurred. Advice should be given to transform the parking lots into green spaces, in order to improve air quality.

The I and L typologies led to wind channeling following the same paths, although parts of the courtyards remain in the "wind shadow" and generate sheltered areas. In the main W-E street, N and NW gradient winds blow from the W, after being channeled between the windward L- and I-shaped buildings. As referred above, the tower-like buildings allow airflow, although causing deviation as the wind diverges upwind of each tower and converges downwind. The experiments with NW wind have shown that the wind speed will decrease not only due to urban roughness, but also to the fact that the wind forms a 45° angle with the prevailing N-S direction of the buildings. When the wind blows from the W direction the existence of the N-S highway (*Eixo Norte-Sul*), which is higher than the surrounding areas, causes eddies near the slopes of the embankment on which the highway is built. The tower-like buildings hinder W wind circulation more effectively than N and NW winds. Wind measurements inside the scale model with Pitot-Prandtl tubes have helped to validate several of the previous experiments. It was shown that the increase of z_0 from 0.02 to 1.5 m in the windward side of Telheiras city district originates a severe drop of near ground wind speed. Wind profiles performed in the scale model showed that when z_0 changes from 0.02 to 1.5 m there is a reduction of 46% in wind speed near the ground, 30% at 50 m high and only 10% at 200 m.

The different experiments show clearly that if the city continues to grow, the N wind circulation will be progressively hindered inside this city district (and others in similar geographical position), with severe negative impacts on inhabitants comfort and health.

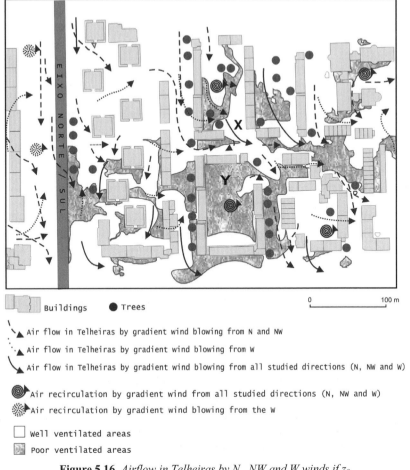

Figure 5.16. *Airflow in Telheiras by N , NW and W winds if z_0 increases to 1.5 m to the N of the city-district [LOP 03]*

5.9. Conclusion

The objective of presenting different approaches and projects was to make clear that in order to study urban climatology, it is not only necessary to monitor urban climate parameters and to use sophisticated models, but also to possess accurate and detailed geographical information, including urban morphology.

Furthermore, accurate measurements require accurate knowledge of the geographical urban environment. Data acquisition is crucial to build and to validate models, but any model may be seriously jeopardized by incorrect measurements. The site of meteorological stations used must be precisely known and, if necessary, the researchers have to set their own networks. In the ESCOMPTE Project, it was shown that wide use was made of sophisticated ground-base and air and sea-borne instruments.

The aim of this text was not to describe different types of models, but to illustrate several procedures based on Lisbon study cases both at the mesoscale and at the microscale. In the study of the urban mesoclimatic patterns of Lisbon, the empirical (statistical) model has made use of a great number of spatialized geographic variables within a 250 m side grid. The multiple regression technique is easy to apply and allows us to separate the influence of urban and other local geographical factors (independent variables) in the observed urban climate features. Several of these were easy to quantify (altitude, latitude, distance to the river); other variables have required the use and/or processing of satellite data, detailed cartography, and sometimes field surveys (NDVI, percentage of built-up area, mean building height). The modeling of PET has been carried out at the microscale based on a similar procedure. The geometry of the city district had to be thoroughly represented in order i) to construct indices (such as SVF) susceptible of "explaining" spatial variation of meteorological parameter or thermal indices; ii) to use the numerical models of microclimatic simulations, such as the *ENVI-met* and the radiation and bioclimate model such as *RayMan*. Numerous physical representations of the geographical areas studied and at different scales were used (Lisbon and the city-district of Telheiras) for wind modeling at the mesoscale and at the microscale: digital terrain models, land cover, canyon geometry, built density, building height. For the city district study, a scale model with all the details of the geographic features had to be constructed (buildings, trees, etc).

Once the researchers have carried out accurate urban climatologic studies (including monitoring, constructing models, validating models), then it will it be possible to proceed to Oke's last three "research modes" (see above) [OKE 06a]. This will allow climatologic issues to be useful to society either in planning and design purposes (mode 6), or impact assessment, including the cost of urban climate and the "cost of inaction" when facing global and local warming in cities, assessing space cooling needs and costs, as well as energy consumption (mode 7). Finally urban climatologists may act in policies development and application (mode 8).

5.10. Acknowledgements

I am indebted to Henrique Andrade and António Lopes, first PhD students and now colleagues, for their valuable contribution to the research in bio- and urban climatology, for the frequent discussions and exchange of views over these issues, all of which have greatly contributed to the development and dissemination of the above topics, for their attentive and critical review of parts of this text and for the use of some original images. I am grateful to João Vasconcelos for the use of Figure 5.11 and for the recent bibliography on Lisbon breezes, to Teresa Vaz for her valuable help with the bibliography and figures and to Teresa Sutcliffe for English proofreading of this text. Part of this research has been carried out in the frame of the UrbKlim Project (POCI/GEO/61148/2004), financed by FCT and FEDER (Operational Programme for Science and Innovation 2010) and of the project "Climatic Guidelines for Planning in Lisbon", financed by the Municipality of Lisbon. Last but not least, Andreas Matzarakis, consultant of the Urbklim project, must be warmly thanked for his helpful comments on two versions of this chapter.

5.11. Bibliography

[ALC 87] ALCOFORADO M-J., "Brisas estivais do Tejo e do Oceano na região de Lisboa", *Finisterra – Revista Portuguesa de Geografia*, vol. XXII, no. 43, pp. 213-225, 1987.

[ALC 92] ALCOFORADO M-J., "The climate of Lisbon's region: thermal contrasts and rhythms", Memória 15, Centro de Estudos Geográficos, Lisbon, 1992.

[ALC 94] ALCOFORADO M-J., "L'extrapolation spatiale de données thermiques en milieu urbain", *Publications de l'Association Internationale de Climatologie*, vol. 7, pp. 493-502, 1994.

[ALC 05] ALCOFORADO M-J., LOPES A., ANDRADE H., VASCONCELOS J., VIEIRA R., *Orientações climáticas para o ordenamento em Lisboa*. Relatório da Área de Geo-Ecologia, nº 4, Centro de Estudos Geográficos, Lisboa, 2005. http://pdm.cm-lisboa.pt/pdf/RPDMLisboa_avaliacao_climatica.pdf

[ALC 06a] ALCOFORADO M-J., ANDRADE H., "Nocturnal urban heat island in Lisbon (Portugal): main features and modelling attempts", *Theoret. Appl. Climatol.*, vol. 84, no. 1–3, pp. 151-159, 2006.

[ALC 06b] ALCOFORADO M-J., ANDRADE H., LOPES A., VASCONCELOS J., VIEIRA R., "Observational studies on summer winds in Lisbon (Portugal) and their influence on daytime regional and urban thermal patterns", *Merhavim*, vol. 6, pp. 90-112, 2006.

[ALC 07] ALCOFORADO M-J., ANDRADE H., LOPES A., OLIVEIRA S., "A ilha de calor de Lisboa. Aquisição de dados e primeiros resultados estatísticos para aplicação ao ordenamento urbano". *Geophilia – o sentir e os sentidos da geografia*, Centro de Estudos Geográficos, Lisboa, p.593-612, 2007.

[ALC 08] ALCOFORADO M-J., ANDRADE H., "Global warming and urban heat island", in MARZLUFF J.M., SHULENBERGER E., ENDLICHER W., ALBERTI M., BRADLEY G., RYAN C., SIMON U., ZUMBRUNNEN C., *Urban Ecology, An International Perspective on the Interaction Between Humans and Nature*, Springer, pp. 249-262, 2008.

[ALC 09] ALCOFORADO M-J., ANDRADE H., LOPES A., VASCONCELOS J., "Application of Climatic Guidelines to Urban Planning. The example of Lisbon (Portugal)", *Landsc. Urban Plan.*, vol. 90, pp. 56-65, 2009.

[AND 98] ANDRADE H., LOPES A., "Utilização de um SIG para a estimação das temperaturas em Lisboa", in FERNANDEZ GARCIA F., GALÁN GALLEGO E., CANADA TORRECILLA R. (coord.), *Clima y ambiente urbano en cidades ibéricas e iberoamericanas*. Madrid, Editorial Parteluz, pp. 85-91, 1998.

[AND 03] ANDRADE H., "Human bioclimate and air temperature in Lisbon", PhD thesis, University of Lisbon, 2003.

[AND 05] ANDRADE H., "O clima urbano – natureza, escalas de análise e aplicabilidade", *Finisterra – Revista Portuguesa de Geografia*, vol. XL, no. 80, pp. 67-91, 2005.

[AND 08] ANDRADE H., ALCOFORADO M-J., "Microclimatic variation of thermal comfort in a district of Lisbon (Telheiras) at night", *Theoret. Appl. Climatol.*, vol. 92, no. 3-4, pp. 225-237, 2008.

[ARN 03] ARNFIELD A. J., "Two decades of urban climate research: a review of turbulence, exchanges of energy and water, and the urban heat island", *Int. J. Climatol*, vol. 23, no. 1, pp. 1-26, 2003.

[ATK 74] ATKINSON B.W., "The reality of the urban effect on precipitation. A case-study approach", in *Urban Climates, WMO Technical Note*, 108, WMO, Geneva, p. 342-60, 1974.

[BAK 08] BAKLANOV, A., GRIMMOND, S., MAHURA, A., ATHANASSIADOU, M., "Enhancing mesoscale meteorological modelling capabilities for air pollution and dispersion applications", COST Action, Report 728, 2008.

[BAT 06] BATCHVAROVA E., GRYNING S.-E., "Progress in urban dispersion studies", *Theoret. Appl. Climatol.*, vol. 84, no. 1-3, pp. 57-67, 2006.

[BRA 05] BRÁZDIL R., PFISTER C., WANNER H., v. STORCH H., LUTERBACHER J., "Historical climatology in Europe – the state of the art", *Clim. Change,* vol. 70, pp. 363-430, 2005.

[BRI 06] BRIDGMAN H.A., OLIVER J.E., *The Global Climate system: Patterns, Processes and Teleconnections*, Cambridge University Press, 2006.

[BOR 00] BORNSTEIN R., LIN Q., "Urban heat islands and summertime convective thunderstorms in Atlanta: three case studies", *Atmos. Environ.*, vol. 34, pp. 507-516, 2000.

[BRU 98] BRUSE M., FLEER H., "On the simulation of surface plant air interactions inside urban environments", *Environ. Model. Softw.*, vol. 13, pp. 373–384, 1998.

[BRU 99] BRUSE M., "Modelling and strategies for improved urban climates", *Proceedings of the 15th International Congress of Biometeorology & International Conference on Urban Climatology*, Sydney, Australia, Macquarie University, 1999.

[CAM 02] CAMUFFO D., "History of the long series of daily air temperature in Padova (1725-1998) ", *Clim. Change*, vol. 53, pp. 49-65, 2002.

[CAR 92] CARREGA P., "Topoclimatologie et habitat", PhD thesis, University of Nice, 1992.

[CAR 98] CARREGA P., "Les spécificités de l'îlot de chaleur urbain à Nice", *Nimbus*, no.13-14, Societa Meteorologica Subalpina, Torino, pp. 33-41, 1998.

[CER 95] CERMAK J.E., DAVENPORT A., PLATE E., VIEGAS D.X. (ed.), *Wind Climate in Cities*, NATO ASI. Series E: Applied Sciences, Kluwer Academic Publishers, 1995.

[CHA 74] CHANGNON S.A., ATKINSON B.W., "Recent studies of urban effects on precipitation in the United States", in *Urban Climates, WMO Technical Note*, vol. 108, WMO, Geneva, pp. 327-44, 1974.

[DET 74] DETTWILLER J., "Incidence possible de l'activité industrielle sur les précipitations à Paris", in *Urban Climates, WMO Technical Note*, 108, WMO, Geneva, pp. 363-364, 1974.

[FEZ 95] FEZER F., *Das Klima der Städte*, Klett-Perthes Verlag, 1995.

[GAR 99] GARCIA E., CARREGA P., "Topoclimatologie urbaine: l'exemple de la ville de Marseille", *Publications de l'Association Internationale de Climatologie*, vol. 11, pp. 424-432, 1999.

[GRI 06a] GRIMMOND S., "Variability of urban climates", in BRIDGMAN H.A, OLIVER J.E., *The Global Climate System. Patterns, Processes and Teleconnections*, Cambridge University Press, p. 210-223, 2006.

[GRI 06b] GRIMMOND S., "Progress in measuring and observing the urban atmosphere", *Theoret. Appl. Climatol.*, vol. 84, no. 1-3, pp. 3-22, 2006.

[HEL 99] HELBIG A., BAUMÜLLER J., KERSCHGENS M-J., *Stadtklima und Luftreinhaltung*. 2., Vollständig Überarbeitete und Ergänzte Auflage. Berlin, Springer-Verlag, 1999.

[HOP 99] HÖPPE P., "The physiologically equivalent temperature–a universal index for the biometeorological assessment of the thermal environment", *Int. J.Biometeorol.*, vol. 43, pp. 71–75, 1999.

[JAU 97] JAUREGUI E., "Climates of tropical and subtropical cities", in YOSHINOM M., DOMRÖS M., DOUGUÉDROIT A., PASZYNSKI J., NKEMDIRIM L., *Climates and Societies – a Climatological Perspective. The Geojournal Library*. Kluwer Academic Publishers, Dordecht, pp. 361-373, 1997.

[JOL 96] JOLY F., "Modélisations à grande échelle de la variation spatiale des températures", *Publications de l'Association Internationale de Climatologie*, vol. 9, pp. 219-227, 1996.

[KAN 06] KANDA M., "Progress in the scale modeling of urban climate: review", *Theoret. Appl. Climatol.*, vol. 84, no. 1-3, pp. 23-33, 2006.

[KRA 37] KRATZER A, *Das Stadtklima*, Vieweg und Sohn, Braunschweig (1st edition), 1937.

[KUT 04] KUTTLER W., "Stadtklima, Teil 1: Grundzüge und Ursachen", *UWSF - Zeitschrift für Umweltchemie und Öko-toxikologie*, vol. 16, no. 3, pp. 187-199, 2004.

[LAN 81] LANDSBERG H., *The Urban Climate*, International Geophysics Series, vol. 28, Academic Press, New York.

[LEG 92] LEGRAND J.P., LEGOFF M., *Les observations météorologiques de Louis Morin*. Monographie 6, Météorologie Nationale, Paris, 1992.

[LIP 82] LIPPS F.B., HEMLER R.S., "A scale analysis of deep moist convection and some related numerical calculations", *J. Atmos. Sci.*, vol.39, pp. 2192-2210, 1982.

[LOP 01] LOPES A., VIEIRA H., "Heat fluxes from Landsat images: a contribution to Lisbon urban planning", *Regensburg Geographische Schriften*, vol. 35, pp. 169-176, 2001.

[LOP 02] LOPES A., "The influence of the growth of Lisbon on summer wind fields and its environmental implications", *Proceedings of the Tyndall/CIB International Conference on Climate Change and the Built Environment*. Manchester, 2002.

[LOP 03] LOPES A., "Changes in Lisbon's urban climate as a consequence of urban growth. Wind, surface UHI and energy budget", PhD thesis, University of Lisbon, 2003.

[LOP SUB] LOPES A., SARAIVA J., ALCOFORADO M-J., "Numerical modelling and wind tunnel experiments to assess wind climate modifications due to urban growth and its environmental consequences", *Environ. Model. Softw.*, submitted.

[LOW 77] LOWRY W.P., "Empirical estimation of urban effects on climate: a problem analysis", *J. Appl. Meteorol.*, vol. 16, pp. 129-153, 1977.

[LOW 98] LOWRY W.P., "Urban effects on precipitation amount", *Prog. Phys. Geogr.*, vol. 22, pp. 477-520, 1998.

[MAN 74] MANLEY G., "Central England temperatures: monthly means 1659 to 1973", *Q. J. Roy. Meteorol. Soc.*, vol. 100, pp. 389-405, 1974.

[MAR 07] MARTILLI A., "Current research and future challenges in urban mesoscale modeling", *Int. J. Climatol.*, vol. 27, no. 14, pp. 1909-1918, 2007.

[MAS 06] MASSON V., "Urban surface modelling and the meso-scale impact of cities", *Theoret. Appl. Climatol.*, vol. 84, no. 1-3, pp. 35-45, 2006.

[MAT 99] MATZARAKIS A., MAYER H., IZIOMOM E, "Applications of a universal thermal index: physiologically equivalent temperature", *Int. J. Biometeorol.*, vol. 43, pp. 76–84, 1999.

[MAT 00] MATZARAKIS A., RUTZ F., MAYER H., "Estimation and calculation of the mean radiant temperature within urban structures", in DE DEAR R.J., KALMA J.D., OKE T.R., AULICIEMS A. (eds), *Biometeorology and Urban Climatology at the Turn of the Millennium: Selected Papers from the Conference ICB-ICUC'99*, Sydney, WCASP-50,WMO/TD, no. 1026, pp. 273–278, 2000.

[MAT 01] MATZARAKIS A., "Die thermische Komponente des Stadtklimas", *Berichte des Meteorologischen Institutres der Universität Freiburg*, vol. 6, 2001.

[MAT 07] MATZARAKIS A., GEORGIADIS T., ROSSI F., "Thermal bioclimate analysis for Europe and Italy". *Il Nuovo Cimento,* C 30, pp. 623-632, 2007.

[MOR 93] MORTENSEN N., LANDBERG L., TROEN I., PETERSEN E., *Wind Atlas Analysis and Application Program (WAsP)* (Vol. I and II), Roskilde, Denmark. Risø National Laboratory, 1993.

[NAK 88] NAKAMURA Y., OKE T.R., "Wind, temperature and stability conditions in an east-west oriented urban canyon", *Atmos. Environ.*, vol. 22, no. 22, pp. 2691-2700, 1988.

[NET 05] NETO J.M.B., "Estudo da circulação atmosférica de Verão sobre a região de Lisboa. Interacção entre a brisa de mar, os efeitos da ilha urbana, a orografia e a presença do estuário", MSc thesis, University of Évora, 2005.

[OKE 87] OKE T.R., *Boundary Layer Climates*, Methuen, London, 1987.

[OKE 88] OKE T.R., "The urban energy balance", *Prog. Phys. Geogr.*, vol. 12, no. 4, pp. 471-508, 1988.

[OKE 06a] OKE T.R., "Towards better communication in urban climate", *Theoret. Appl. Climatol.*, vol. 84, no. 1-3, pp. 179-190, 2006.

[OKE 06b] OKE T.R., "Initial guidance to obtain representative meteorological observations at urban sites", *Instruments and Observing Methods*, report no. 81WMO/TD-No. 1250, World Meteorological Organization, 2006.

[OLI 07] OLIVEIRA S., ANDRADE H., "An initial assessment of the bioclimatic comfort in an outdoor public space in Lisbon", *Int. J. Biometeorol.*, vol. 52, no.1, pp. 69-84, 2007.

[OOK 07] OOKA R., "Recent development of assessment tools for urban climate and heat-island investigation especially based on experiences in Japan". *Int. J. Climatol.,* vol. 27, no. 14, pp. 1919-1930, 2007.

[PAR 98] PARLOW E., "Analyse von Stadtklima mit Methoden der Fernerkundung", *Geographische Rundschau*, vol. 50, no. 2, pp. 89-93, 1998.

[PEA 07] PEARLMUTTER D., BERLINER P., SHAVIV E., "Urban Climatology in arid regions: current research in the Negev Desert", *Int. J. Climatol.*, vol. 27, pp. 1875-1885, 2007.

[PET 97] PETERSEN E., MORTENSEN N., LANDBERG L., HØJSTRUP J., FRANK H., *Wind power Meteorology*, Roskilde, Risø National Laboratory, 1997.

[PFI 94] PFISTER C., BAREISS W., "The climate in Paris between 1675 and 1715 according to the meteorological journal of Louis Morin", in FRENZEL, B., *Climatic Trends and Anomalies in Europe 1675-1715*, G. Fisher, pp.151-171, 1994.

[PRI 01] PRIOR V., CARVALHO R., NETO J., MANSO M.D, "Simulation du vent sur des zones urbaines côtières, abstract presented at the *12th Meeting of the Association Internationale de Climatologie*, Seville, p. 209-210, 2001.

[RAT 06] RATTI C., DI SABATINO S., BRITTER R., "Urban texture analysis with image processing techniques: winds and dispersion", *Theoret. Appl. Climatol.*, vol. 84, no. 1-3, pp. 77-90, 2006.

[RIG 08] RIGBY M., TOUMI R., "London air pollution climatology: Indirect evidence for urban boundary layer height and wind speed enhancement", *Atmos. Environ.*, vol. 42, pp. 4932-4947, 2008.

[SAI 04] SAILOR D.J., LU L., "A top-down methodology for developing diurnal and seasonal anthropogenic heating profiles for urban areas", *Atmos. Environ.*, vol. 38, no. 17, pp. 2737-2748, 2004.

[SHE 05] SHEPHERD J.M., "A review of current investigations of urban-induced rainfall and recommendations for the future", *Earth Interact.*, vol. 9, n° 12, p. 1-27, 2005.

[SIM 94] SIMPSON J.E., *Sea Breeze and Local Wind*, Cambridge University Press, Cambridge, 1994.

[SVE 02] SVENSON M, ELIASSON I, "Diurnal air temperatures in built-up areas in relation to urban planning". *Landsc. Urban Plan.*, vol. 61, pp. 37–54, 2002.

[VAS 04a] VASCONCELOS J., "Avaliação climática para o planeamento urbano de Lisboa: influência do crescimento urbano no sistema de brisas do estuário do Tejo", MSc thesis, New University of Lisbon, 2004.

[VAS 04b] VASCONCELOS J, LOPES A., SALGADO R, NETO J., "Modelling of the estuarine breeze of Lisbon (Portugal): preliminary results", in García Codron, J.C. *et al.*, ed., *El Clima entre el Mar y la Montaña.* Asociación Española de Climatología y Universidad de Cantabria, Serie A, n° 4, Santander, p. 165-169, 2004 http://www.aeclim.org/4congr/vasconcelosJ04.pdf

[WIL 95] WILKS D.S., *Statistical Methods in the Atmospheric Sciences*, Academic Press, San Diego, 1995.

[WMO 96] WORLD METEOROLOGICAL ORGANIZATION, *Guide to Meteorological Instruments and Methods of Observation*, 6th edition, WMO-No. 8, Geneva, 1996.

[YOS 90/91] YOSHINO M.M, "Development of urban climatology and problems today", *Energy Buildings*, vol. 15-16, pp. 1-10, 1990/91.

Chapter 6

Geographical Information, Climate and Atmospheric Pollution

6.1. Introduction

Climate and atmospheric pollution are two atmospheric hazards that are very closely related. Pollution levels are not uniform throughout the world, and this is particularly true for pollution that is created by human activity. However, it is the concentration of pollution in the atmosphere which is dangerous to human health, and which is also damaging the biosphere. This concentration of pollution in the atmosphere varies, and depends on different meteorological conditions that influence the way in which the pollutants are dispersed into the Earth's atmosphere. Strong winds help the pollutants to disperse, whereas more stable conditions, such as thermal inversions, act as a lid trapping the pollutants that build up on the ground. However, the relationship between climate and pollution has experienced many changes throughout the years: the days when acidic particle pollution only occurred during winter anticyclones, associated with long periods of cold, dry weather, have long gone. The time of year when the most pollution is present in the atmosphere is during long, hot, sunny days in summer. These weather conditions are associated with ozone levels that are harmful to our health. Whether it be summer or winter, pollution levels in the atmosphere depend on changes in weather conditions.

The geographical information that is available on these issues provides much more information than simple pollution measurements. This is true despite the progress that has been made in metrology; progress that has influenced monitoring strategies. The information is formatted and presented according to the expectations of our society. In the age of industrial pollution there was a culture of secrecy that prevailed over the obligation to publish information on pollution levels. This obligation to provide information relating to pollution levels is now part of the

---

Chapter written by Isabelle ROUSSEL.

French Environmental Charter, which has had Government backing since 2004[1]. Geographical information is a domain in which the physical measurements, which are created by very technical cognitive methods, can have an effect on society as far as developments and social responses are concerned. This method of investigation adds to the problem of atmospheric pollution, which is measured in terms of its damaging effects or its harmful impacts.

With this in mind, geographical information depends on the world of metrology and the world of politics. A few years ago the aim of the information that was to be published was to forecast pollution levels that would damage the quality of air. For a long time, the word alert was the key word that was used to provide information on atmospheric pollution in an avoidance strategy. As years went by the relationship between climate and pollution became much more complex both in spatial and temporal terms. The pollutants emitted, even if they had dispersed into the air, were able to change their chemical make-up and remain in the atmosphere, which was no longer seen as a container of chemically neutral elements but as a laboratory in which a real alchemy was taking place. The pollutants in the atmosphere no longer only made people cough but were beginning to weaken the entire equilibrium of the planet's climate. Geographical information no longer only dealt with short-term periods of peak pollution, but could also be applied to longer time periods, in other words climate change.

Furthermore, information that was available on pollution levels no longer only provided information relating to the measurement site, but started to provide information relating to the region in which the measurement site was located. It is these successive changes in scale that are also responsible for the changes that have been made to the geographical information available. By definition, geographical information alters scales and connects one scale to another. The different types of information available have evolved according to metrology: from analyzers, which provided information on global acidity levels on a daily or weekly basis, to the current systems of ion chromatography, which are used to detect the nanograms or picograms present in polluting gases. Specific measurements are discretized and different models make it possible to plot these measurements on a map thanks to the use of digital applications or deterministic simulations. Periods of local peak pollution and local alerts are forecast on a regional level meaning that at the moment there is a transition from the local level to a more global level. LAURE[2], with the

---

1 Article 7 from the French Environment Charter, "Everyone has the right, under conditions and limits that are defined by the law, to access information concerning the environment which is held by public authorities. Everyone also has the right to participate in the decision making process when it comes to developing ideas which have an impact on the environment".

2 LAURE French law, created in December 1996, on air and the rational use of energy.

help of several regional policies that were adopted in France[3], have meant that the management of atmospheric pollution can now be considered on a regional level. The idea to regionalize atmospheric pollution management is justified by administrative and political needs as the prevention of atmospheric pollution depends on many factors that are managed at regional levels, such as transport, habitat, energy use, etc. However, the possibility of creating new tools that can be used on a regional level raises methodologically complex issues. These issues are associated with the difficult spatialization of pollution levels that are measured in more specific areas. The information gathered is listed in various documents, and conforms to the demands of geographical information. It can then be used to link pollution levels with other factors so that it is possible to identify the tools that can be used to prevent and fight against pollution. This is the advantage of using geographical information systems (GISs).

The creation of these highly technical documents also raises the issue of how they are used, and questions the relationship between the science and geographicalal information that is published in them. The information that is produced in these documents is undoubtedly produced by experts who work in the field of pollution, but the information is also used as a power tool and a communication tool by a shared government. The interface between technical knowledge and the reasons why this information is used is becoming more and more complex. The almost linear nature of the relationship between emissions and immissions when an alert for high pollution levels is given means that the regional management of pollution is a much more difficult task. This is due to the fact that the area in which the pollution occurs, as well as the number of pollutants, increase rapidly in size. This issue from the feedback that is created by the information produced and the action that takes place: is it pollution management on a regional level that leads to the creation of tools to help prevent pollution, or is it the information produced that guides and orients the regional management of pollution?

6.2. Peak pollution periods and alerts: pollution and climate paroxysm

Measuring pollution levels and evaluating their impact have changed dramatically over the last few years. First, the linearity that exists between industrial emissions of pollution and air quality have made it possible to predict which weather conditions lead to high pollution levels, for which an alert can be made. However, the confidential characteristics of these alerts, which were developed during a one-off conference held between the Regional Departments for Industry, Research and the Environment in France (DRIRE) and industrialists, makes us question if an alert can remain secret. The nature of atmospheric pollution has also changed, and the way in which it is measured and analyzed is no longer a question of monitoring the large polluting industries. The total amount of pollution produced

---

3 PRQA: French regional policy on air quality; PPA: French policy on atmospheric protection for cities with more than 250,000 inhabitants; and PDU: French policy on urban transport for cities with more than 100,000 inhabitants.

from car exhaust pipes contributes largely to air quality. LAURE has become an essential tool for rationalizing these profound changes. This piece of legislation has defined what is known as measurement networks and has set alert thresholds, which makes it possible to initiate a protocol of media coverage should the pollution levels exceed these alert thresholds.

6.2.1. *Industrial pollution: the era of secrecy*

The first measurements of air quality were taken for hygienic purposes. Deadly levels of acid particle pollution in London in 1952 and all along the Meuse Valley in 1930 caught the attention of health authorities. The health authorities found that prosperous, yet polluting, industrial areas were damaging the air and thus the health of the people living in these areas. It was at this moment that the existence of invisible physical and chemical pollutants were discovered and, as a result, this led to the creation of new pollution management strategies such as the Clean Air Act which was introduced in England[4].

Since the 1960s monitoring is the institutional answer for pollution prevention. This type of monitoring has made industries aware of the harm that their industrial waste can cause. The data gathered in relation to pollution levels makes it possible to carry out studies and to work out guideline levels of pollutants that should be present in the air, which can be used as a basis for the European Environmental Policy [FES 97]. However, the influence that these measurements have on pollution management, as well as the importance of metrology has meant that air and water management is slowly being transferred from hospital doctors to engineers. This change in monitoring features, which move from the local hygiene field to the world of engineers [VLA 99], can also be represented by the creation of a technical network managed by the French state. This evolution is still continuing today with the regionalization of the technical networks and, according to P. Richert [RIC 07], is seen as the result of a need for more specialized and more sophisticated networks, impossible to achieve in small structures. With the creation of the Ministry for the Environment in 1971, and due to the fact that the École des Mines in Paris (a prestigious French engineering school in Paris) controlled how air quality was monitored, monitoring air quality became an issue that was no longer dealt with by health experts, but became part of engineering sciences. As a result of the progress made in metrology, the normal values of pollutants in the air can be broken down into more precise time periods, meaning that it has become easier to understand and predict periods of peak pollution and also to avoid them. The idea of creating an alert is associated with measurements that are recorded continuously, and which are used to detect the harmful risks associated with pollutants.

---

4 The pollution problem in London is an extremely old problem that dates back to before the 17th century. Any problems that existed before this time are also linked to problems associated with pollution after the 17th century. Legislations were passed during the 19th century, in 1853, and in 1863 with the Alkali Act. The Clean Air Act was introduced in 1953.

At the time of industrial pollution the relationship between pollutant emissions and air quality could be described as being linear. Due to the concerns that existed at the time, most of the attention was focused on periods of peak pollution in order to try and limit the amount of industrial emissions during these periods so as not to exceed the threshold level of pollutants present in the air, as allowed in the European Environmental policy. At the time, acid particle pollution was much higher than it is today, and it became necessary to remove the peak pollution periods in areas which were vulnerable to such pollution. These areas included areas in France such as the Étang de Berre, Dunkirk, the lower-Seine, Grenoble, Lyon, etc. Over the last 20 years many different devices have been successful at removing these peak pollution periods. The devices were drawn up by the DRIRE and different industrialists, and also became an integral part of different prefectoral laws. The result of such actions is generally positive, and in order to achieve such a result it is necessary to rely on the use of basic weather forecasting. Measurement tools, however, make it possible to continuously monitor pollution and to easily detect areas of peak pollution.

The aim of these alerts is to remove any periods of peak pollution. In order to achieve this, the alerts serve as a way to force industries into using low sulfur fuels. The alerts, which are operated by the prefects and which are not exposed to much media coverage, rely on the principle of improving air quality by reducing emissions of acid particle pollution. An alert is raised by the DRIRE, based on advice given from meteorologists who rely on weather forecasts. Local negotiations, which involve prefectoral services, measurement networks and industrialists, try to decide on the device that is to be used to help remove the periods of peak pollution. The creation of such devices is based on two main indicators: the first is predictive (forecasting a high-risk weather situation); the second is optional, and refers to the level of air quality. In industrial areas, models that forecast the levels of sulfur pollution are used. These models are relatively easy to use once factors such as temperature, atmospheric pressure, absence of wind, and the existence of a temperature inversion in the lower layers of the atmosphere are known. The alert is preventative, and it can be preceded by a pre-alert during which industrialists are told to pay attention to the amount of pollution that is being produced by their industries.

The use of such a device has come into question with the emergence of car pollution. Preventing car pollution is a much more complicated process. Air quality management, which focuses on monitoring industrial emissions, has changed dramatically with the introduction of many other sources of pollution. At the time when heavy industry in towns started to decline, atmospheric pollution from cars started to increase and became part of a renewed vision of the urban environment, urban ecology and the sustainable city.

6.2.2. *The emergence of car pollution and the systematic creation of a measurement network*

Pollution has changed. There are fewer cases with generalized pollution even if some local factories plumes are still being observed. Figure 6.1 shows how sulfur pollution levels have decreased in an industrial region such as Dunkirk in the North of France, and also shows that car pollution levels are high in the centers of larger towns and cities. Nitrogen dioxide levels, which are considered as an indicator of urban pollution, have not changed very much. The difference between values measured at a background station, and those measured at a near station (see below) are increasing.

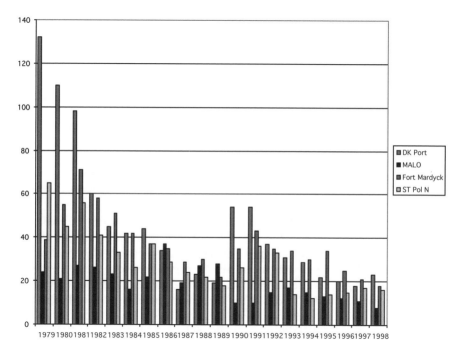

Figure 6.1. *The evolution of annual averages of sulfur dioxide levels for four stations based in Dunkirk, France. The averages of the stations found closest to the factories have also decreased significantly over a period of 20 years. Source: ATMO Nord-Pas-de-Calais*

This evolution of atmospheric pollution has affected Europe as a whole ever since the phenomenon of acid rain highlighted the importance of cross-border pollution. With the introduction of LAURE, France has tried to respond to the different directives that were set by the European Union, by systematizing the monitoring system to be adopted throughout the whole of the country. This task of monitoring pollution levels in France was confined to registered associations that monitor air quality (AASQA). These associations are organizations developed and acknowledged by the French Minister for the Environment and are made up of following the model of associations issued from 1901 law. Information relating to air quality corresponds to a legislative obligation defined by LAURE (30-12-1006). This law is applied to all French cities, and to the models used in the regions of Alsace and in Ile-de-France. The different monitoring associations that exist today, such as the AASQA, are traditional associations that measure industrial pollution and which have been systematized since the introduction of the 1961 law.

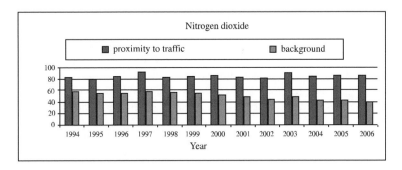

Figure 6.2. *The evolution of annual averages of nitrogen dioxide levels for one background station and one near station (close to a main roadway) in the region of Paris. Source: AIRPARIF*

6.2.2.1. *The first measurement networks follow the hygienism tradition*

The reduction of emissions needs to be explained in more detail by analyzing the impact of pollutants on human health. Local communities became more and more interested in issues relating to air pollution and the first networks that were developed to monitor pollution levels could often be found in large municipal hygiene offices, such as those in Nancy, Lyon, St Etienne, etc. With the creation of these offices it became possible to group together the different people who would be involved in the monitoring process, region by region. The different people involved in the monitoring process included doctors, politicians, universities and industrialists and by grouping these people by region it became possible to create scientific think-tanks on a local level.

This first monitoring phase formed part of many private and local initiatives that were developed to help improve the health of the population. As a result of this first monitoring phase it became possible to develop epidemiological studies that established a clear link between air quality and the health of the population.

Gradually these networks became part of new registered associations that were created by the French Environment Minister. The creation of the Air Quality Agency (AQA) in France in 1981 as well as the generalization of the eco-tax in 1985 made it possible to equip the networks with better performing technology that could be used to improve the quality of physical and chemical measurements recorded in the atmosphere. This new approach, which was adopted by the French

Ministry for the Environment, meant that the Ministry was starting to focus less on health, and more on the polluting industries with the introduction of the DRIRE in the field of monitor air quality.

It was during the time of the rapid increase in urban pollution (linked to the increase in car pollution) that industrial taxes had to be paid to the French government. These taxes would be paid to the State and would then be used to finance the monitoring of air quality in urban areas.

Urban pollution management progressively became part of city politics as time went on. With this in mind more and more city dwellers became interested in information that was published, and which informed them of pollution levels in their own cities. This proves that the culture of secrecy, which was so often linked to periods of industrial development, can no longer be applied to today's society. Today's society is made up of people who use urban transport and includes people who experts in this field wanted to target in order to make them aware of the phenomenon of urban pollution. However, a certain period of time was needed in order to convince car manufacturers and car users of the inadequacy of the cars adaption to the town.

6.2.2.2. *The monitoring system generalized by LAURE*

The network of pollution management associations which monitor air quality (AASQA)[5], has experienced great changes, particularly in relation to their structure and to the way in which they carry out their business. Measuring pollution levels and producing information on the recorded values are subject to the decree that was passed in 1998 and to the law that was passed on March 17, 2003, which relates to public awareness and the way in which air quality is monitored. Unlike hygiene management, monitoring air pollution is expensive for local communities, and producing clear information on pollution levels is obligatory. Publishing the results from pollution analysis is one of the obligations that must be carried out by the measurement networks.

These associations, as in 1901 law, function as a quadripartite organization, in other words it is divided into four key areas. Their board of administrators is divided into four different boards with each board representing the State, the local communities, the industrialists, and the associations themselves[6], respectively. In theory, the first three boards which have been mentioned provide equal amounts of funding to this structure. According to the polluter-pays principle, it is the industrialists who pay part of the eco-tax on atmospheric pollution. In 1999 this license became known as the General Tax on Polluting Activity (TGAP[7]). The industrialists are able to pay part of this tax directly to the network of pollution management associations located closest to the industries in question. The aim of

---

5 Registered association for monitoring air quality.
6 Environmental associations, consumer and qualified persons associations.
7 General Tax on Polluting Activities.

this is to promote air quality monitoring. Despite the increase in the number of taxed pollutants present in the atmosphere (volatile organic compounds and particles that are added to original pollutants to form new ones), and the continual increase in the rate of eco-tax, the tax base has continued to decrease in value. This is due to the fact that the level of industrial pollutant emissions produced has been decreasing, and this is linked to changes in industry. Since 1996, the French government has provided funding to these pollution management associations thanks to LAURE. However, this provision of funding hides the fact that there has been an inevitable decrease in the number of industrialists found to promote air quality monitoring. In addition, local communities are also subsidizing the measurement networks by providing services that extend beyond the strict monitoring obligations that were set on a European level.

6.2.2.3. The AASQA have four main objectives: measuring, informing, alerting, and studying

Management associations need to be aware of the fact that it is difficult to determine the spatial representation of a measurement site when they are developing a network of analyzers that are to be used to measure pollution levels. Over the past few years it has become possible to divide these stations into five different subgroups: background urban stations; stations located close to roadways; stations located close to industries; rural stations used as references; and stations that provide detailed information recorded over a period of time.

The measurements recorded are considered to be indicators of either background or sources of near pollution. A measurement network fluctuates between two functions: the first one involves monitoring the sources of emission, and the second function involves monitoring air quality in densely populated areas or in areas where the population is made up of lots of children, older people, or sick people.

Analyzers used to monitor industrial pollutant emissions have, for a long time, been the most widely available. The low level of sulfur pollution that exists nowadays has meant that each measurement network has had to be restructured so that it can focus on measuring the pollution levels of more worrying pollutants as well as fighting against them. The installation and use of these analyzers is often accompanied by recording mobile measurements; for example by using mobile labs in lorries or by carrying out experiments that make it possible to detect the presence of new pollutants, such as volatile organic compounds, benzene, and pesticides, etc.

The measurement networks need to provide up-to-date information on air quality so that different values can be compared with limit values actually existing for; sulfur dioxide, particles, ozone, nitrogen dioxide, and lead particles. Such a policy induces management strategies, which focus on identifying and removing peaks, more than overcoming the background pollution.

Media coverage of high levels of pollution will also lead to the creation of an alert. Such media coverage will also make more and more city dwellers aware of the notion of atmospheric pollution. This means that it is possible to define the technical capabilities of the monitoring networks. The alerts raised by LAURE are not carried out in the same way as the alerts that were previously raised by industrialists.

6.2.3. *The alert system devised by LAURE*

The main objective of raising such alerts is for the media and for health reasons. These alerts will allow people who are sensitive to the different kinds of pollutants to take a certain number of precautions. As soon as an alert level has been exceeded, an alert is then made to this vulnerable population by providing them with the relevant information (Table 6.1).

Documents used for planning purposes are intent on respecting the coherence of two different processes including: the old process of preventative industrial alerts used in licensing actions, and the current process of making information public. The atmosphere protection plan (PPA), must be applied in cities that have a population of more than 250,000 inhabitants. The aim of the PPA is to record the different protocols of alert that are used, and which have already been set out in prefectoral decrees.

The acquired experience in industrial pollution alerts cannot be easily applied to pollution peaks created by car pollution. This is due to the fact that car pollution is produced by many different pollutants that are referred to as secondary pollutants. The reason alerts, which are used to warn people about industrial pollution levels, are so effective is due to the linearity that exists between emissions and emissions. A reduction in the number of pollution sources leads to an improvement in air quality. Reducing the emissions of industrial pollution is an easy task because there are not many different sources of pollution that exist. As far as car pollution is concerned, and especially in the case of ozone, a major reduction in the most harmful pollutants does not mean that a decrease in ozone levels will be recorded on a local level as ozone levels are measured on a continental scale.

The aim of these alerts, introduced by the LAURE, is not operational because they were introduced to make the general public aware of pollution levels. The introduction of these alerts has made the public aware of the fact that nowadays it is no longer the polluting industries that we should be focusing our attention on in order to fight against pollution, but rather that each of us contributes to the pollution of the environment through car and domestic pollution. Media coverage that is associated with high-level pollution, in other words levels that exceed the maximum acceptable threshold, is used as a sort of risk management. Such media coverage has two main objectives: first, to make the population aware of pollution levels, and second, to inform the population about what they can do in order to reduce pollution levels. Paradoxicaly, this short term advice may have some results in the long term.

Different alert thresholds are set, depending on the type of pollutant being monitored (Table 6.1). In certain regions in France particles are included in these alert advices. These thresholds are detected by measurement networks that publish information. The statutory device that is used to prevent such pollution is controlled by the different departments authorities in France. In this type of system air quality management focuses on the pollution paroxysmique episode. The geographical information that is produced as a result of this is, therefore, very specific and geared towards the risk of an exceptional event The alert generates more attention than general information trivialized through the use of the Atmo index. The aim of the Atmo index is to quantify air quality in a geographicalal area using a dimensionless figure generated from data recorded for different types of pollution. These different pollution types are made up of different pollutants recorded at different measuring sites.

POLLUTANTS	THRESHOLDS ($\mu g/m^3$)	
	INFORMATION-RECOMMENDATION	ALERT
Nitrogen dioxide	200	400 or 200, if the procedure of information-recommendation was started the day before and on the same day; and if forecasts lead experts to think that an alert may need to be raised the next day
Sulfur dioxide	300	500, if this value has been exceeded for a period of 3 consecutive hours
Ozone	180	1st threshold is 240 if this value has been exceeded for a period of 3 consecutive hours 2nd threshold is 300 if this value has been exceeded for more than 3 consecutive hours 3rd threshold is 360

Table 6.1. *Information-recommendation thresholds and measures taken to alert the population (decrees from 02/15/2002 and 11/12/2003, and the lawpassed on 08/17/1998)*

The Atmo index, inspired by the American threshold index, is a ratio measured between pollution levels and reference values that are calculated for each pollutant. However, the use of this index is rather limited because it can only be used in areas in which the exact same type of pollution is present. It is not possible to integrate data produced by stations located near the source of pollution into this type of index as this index does not take the quality of the surrounding air into consideration. In addition, the Atmo index only corresponds to the most concentrated pollutant that can be found in the surrounding air. The Atmo index also does not consider the quantity of other pollutants or the influence that these other pollutants have on the quality of the surrounding air.

This index, which can be used as a real communication tool, prevents geographical information from producing accurate data on the actual pollutants present in the air, which could lead to different studies being carried out.

In order to broadcast peak pollution levels effectively, it is necessary to introduce prevention devices for those people who are sensitive to the different pollutants found in the air. It is necessary to be able to predict pollution levels and this is made possible by creating models that can be used for this purpose. As far as global pollution is concerned, e.g. ozone, the prevention and removal of peak pollution periods is difficult to carry out in the short term. As can be seen in the example of the alert that was raised in Paris in October 1997 [MIE 00], there have been so many changes in meteorological conditions are much more effective a decrease in the amount of pollutants emitted on a local scale, this decrease would still not have an effect on pollution levels on a global scale.

Deciding when such alerts should be raised nowadays relies on the effective forecasting of peak pollution levels and, therefore, on the use of different types of meteorological and chemical models used alongside other expert systems. Modeling air quality is necessary and requires the use of spatial information. Territories, as well as their geographical characteristics, contribute to providing lots of information about air quality even over a short time

The information that is available, which relates to atmospheric pollution, has become increasingly complex as time has gone by. The culture of secrecy is no longer applied to today's society, and extensive media coverage of pollution levels acts as an important link used to broadcast the nature of the atmospheric pollution which concerns every body. Analyzing the different alerts carried out shows how insufficient the different measurements that correspond to industrial pollution have actually become. Now, it is not possible to reduce pollution monitoring to comparing the measurements registered with limit values. It is no longer the factories and the large cities that can be singled out for global pollution levels, but global surroundings. Over time, measuring the amount of pollutants present in the atmosphere has become more diverse, especially regarding discovering new pollutants. In addition, there is also a greater density measurement points as they are organized into different measurement networks. Each territory is not only an monitoring subject, through administrative limits. Since the introduction of LAURE back in 1996, there has been a more spatial vision as far as air quality is concerned and this has been made possible thanks to the use of specific measurements that focus on peak pollution periods. Geographical information is changing its focus at the moment, and instead of focusing on time and peak pollution it is now starting to focus on space and measurements.

6.3. LAURE and territory age

For a long time now information used in the field of atmospheric pollution has been focused on the evolution of levels over time for both administrative and health reasons. Conversely, spatial information has become increasingly used because modeling techniques have improved the quality and reliability of the spatial information available.

European directives predict that modeling will be increasingly used as a technique to measure air quality. It will be used alongside the measurements recorded by the measurement networks. The aim of this approach is to examine the level of atmospheric pollutants present in the atmosphere on a global surrounding, even in regions that have low levels of pollution and do not require an analyzer to continuously check whether pollutants are present in the atmosphere. Frameworks drawn up for monitoring air quality describe the different methods used to monitor air quality for a particular country. The law passed on March 17, 2003, concerning the monitoring of air quality as well as the provision of information to the general public, enables each AASQA to evaluate its own monitoring equipment, and to adjust it for any changes that occur in air quality. The equipment can be adjusted thanks to the introduction of an air quality monitoring program (PSQA). In areas that have high levels of pollution or in areas where it is likely that the maximum pollution thresholds (set by the European Commission) will be exceeded, the air is continuously monitored by using sensors and analyzers located at fixed measuring sites. Whenever the pollution level is situated between the maximum and minimum threshold levels, air quality monitoring is carried out by a series of measurement campaigns and modeling. Whenever pollution levels are lower than the minimum threshold level then modeling can be used. Specific studies can be carried out for certain measurement sites, such as those close to large industries or to areas in which there is a lot of traffic.

Nowadays, however, preventing atmospheric pollution does not only involve monitoring air quality, as was the case during the period of heavy industrial pollution. Currently, to prevent atmospheric pollution it is necessary to examine how different sources are managed; these sources include transport, housing, and agriculture, etc. It is possible to use a GIS to respond to these issues as the GIS is capable of linking air quality with other geocoded factors that can be used in three ways: first of all the GIS can be used to identify ways in which atmospheric pollution can be prevented on a local level. It can also be used to evaluate the benefits of projects undertaken monitoring the decrease in pollution levels. Finally, it can be used to examine the health risks created by exposure to the pollutants. This final method is possible by identifying those members of the population who have been exposed the most to the different pollutants present in the atmosphere.

These complex information systems, however, rely on a paradox. There tend to be less spatial variations of atmospheric pollution than variations of atmospheric pollution measured over time. This is due to the influence that meteorological factors have on the concentration of the pollutants present in the atmosphere and on the different levels of pollutants present on a daily basis. The spatialization of air quality is paradoxical for an element which crosses the boundaries [ROU 01b].

The spatialization of air quality assumes that three phases take place. The first phase is inventory a list of sources, as well as working out the spatialization of the sources; the second phase involves plotting the air quality measurements on a map; and the final phase involves linking atmospheric pollution with other geocoded data so that it is possible to evaluate the exposure to pollution and the health risks associated with this.

Developing tools that can be adapted to these three different phases of spatialized atmospheric pollution leads to the creation of some major methodological problems. These problems correspond to the superposition of three different models. One model is used for the spatialization of emissions, whilst another converts emissions to immissions, which is made possible by integrating the difficulties associated with mapping the climate into the model. The remaining model, which is used to measure the spatialization of risks associated with atmospheric pollution, shows that on a local scale it is still quite difficult to measure how much exposure the population has had to atmospheric pollution. The superposition of these three approaches leads to the creation of different problems for each of the different models.

6.3.1. *A spatialized emissions inventory*

If the different pollutants are able to move from country to country and if the sources of pollution are found at a local level, this means that the different sources need to be identified and included in an emission inventory that is created and provides information on the different emitting sources that send pollutants into the atmosphere. These documents, which are created for each pollutant, incorporate the quantities of gas that are emitted by a surface unit whose size is fixed and whose size depends on the area under investigation. The creation of these inventories questions the spatialization of data, and the ventilation of a specific piece of information which can be found in a particular spatial unit. Martinet [MAR 04] based his thesis on these methodological issues. How may the emissions coming from one factory chimney be affected in different spatial units? The spatialized emissions inventories are essential tools that can be used to model and map air quality. They are also tools that can be used to examine whether the pollution level is decreasing/increasing. When these tools are used for greenhouse gases, they can help decrease greenhouse gas emissions by a factor of four. Therefore, it is necessary that the methodologies adopted for creating such inventories all over Europe are uniform, and that these inventories are scientifically tested and verified. With the use of its Corin-air data, the European Union is developing a coherent set of European inventories and is working on harmonizing the different emission factors that are included in these inventories. In France, the Technical

Interprofessional Center for Atmospheric Pollution (CITEPA[8]), which is based in Paris, monitors and spatializes pollution emissions.

The creation of these land inventories relies on the use of a census created on an annual basis or over a shorter period of time. This register is in an essential tool that can be used to map and model the spatial distribution of emissions and pollution levels.

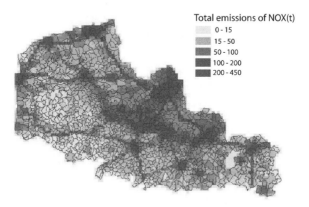

Figure 6.3. *Example of a land inventory showing emissions of pollution in the French region of Nord-Pas de Calais: NO$_X$ emissions produced by vehicles, for a grid of 2×2 km² (traffic data 2002) Source: [MAR 04] (see color section)*

6.3.2. *Air quality maps and models*

There are many different ways in which air quality maps can be created. These maps show the spatial variations of pollution levels for a particular area. They can be created as the result of a simple process of interpolation, which is applied to data that has been recorded by a dense measurement network. They can also be the result of a more sophisticated interpolation process in which lots of data are integrated, or they can be the result of an air quality model.

6.3.2.1. *Maps and air quality*

The distribution of pollution between two measurement points does not only depend on distance if there are other sources of pollution found between the two points:

– the use of passive tubes (available at a moderate cost) leads to the creation of a strong density of measurement points influenced by all sources of pollution that are present in a particular area. When the area being studied is not very large, it is possible to use traditional spatial interpolation methods, such as kriging, to create a satisfactory representation of the spatial distribution of air pollution. The results from these campaigns and these types of maps are available in numerous regions in

8 http://www.citepa.org/.

France for ozone, nitrogen oxide, and benzene. For example, the ozone map for the region of Rhône-Alpes is very similar to the region's relief map. The ozone map shows the presence of high levels of ozone, which were recorded at measurement sites located at higher altitudes within the region. The measurement techniques using passive tubes accumulate the different pollutants over a period of at least 1 week, and with this in mind, it becomes very difficult to detect peak pollution periods. These maps show the presence of background pollution, which is an indicator of a high exposure to pollutants. In order to be more representative, the measurements need to be carried out under different weather conditions, e.g. in summer and winter, during calm conditions, and during stable, anticyclone conditions, etc;

– the use of analyzers makes it possible to store the information that is gathered with a small time step; however, not enough analyzers are used so that a simple process of spatial interpolation can take place. Therefore, it is necessary to use data from emission inventories with the aim of integrating additional sources of pollution into the interpolation process. This technique for creating air quality maps is now commonly used.

6.3.2.2. *Modeling air quality*

There is a lack of measurement sites available that provide information on the slight differences in pollution caused by different topoclimatic factors. This is true for areas in which pollution levels are not systematically measured and for small-scale maps that provide information for a limited zone. In this type of situation it is necessary to model the information relating to air quality that has been gathered.

Air quality depends on different meteorological factors found at many different and imbricated scales, which control the dispersion of pollution emitted from the pollution's sources. Modeling air quality adds to the complexity of the evolution of different chemical processes to the atmospheric dynamic. The models, which are seen as being deterministic, take the spatialization of air quality into consideration; however, they cannot get rid of the uncertainty linked with the instability of the climate.

The climate system is not an entirely predictable system, and this notion of unpredictability is linked to a mathematical property associated with the equations of displacement. These equations are not linear and they confuse scales of time and space. This is the reason why the equations cannot be solved without a computer. It is only possible to predict what is going to happen in the atmosphere for a period of approximately 10 days. Beyond this timescale it is not possible to provide an accurate forecast due to the unstable characteristics of the weather conditions. This well-known effect, known as the butterfly effect, was developed by Edward Lorenz in 1963. The butterfly effect states that any degradation in weather conditions, irrespective of how small this degradation might be, will irreversibly change the history of the atmosphere [TRE 06]. This notion of uncertainty increases with the spatial resolution of air quality. Modeling seemed more suited to the spread of large-scale phenomena, while at a local scale the numerous factors, which affected air quality, raised uncertainty. So the spatial representation given by background is better than this given by stations located near the source of pollution. In general, all of these models can be used to create maps that provide information on air quality.

According to different definitions[9], atmospheric pollution is not evaluated in terms of how chemically (or biologically) pure the atmosphere is but rather in relation to the damaging consequences that atmospheric pollution has on humans and on human health. The tools necessary for air quality management need more than measurements data that provides information on the concentration of different pollutants present in the atmosphere. The tools used need to be used in conjunction with investigations carried out into the effects of pollution on human health, as well as the effects of pollution on vegetation and on the planet as a whole. This essential link in the cognitive chain has its own uncertainty and vulnerability and makes the relationship between knowledge and action in the field of atmospheric pollution more complex.

6.3.3. *Evaluating and mapping the impacts of pollution in order to improve prevention*

The use of GISs provides an answer to the problem of atmospheric pollution, which is defined by its harmful effects. Therefore, in order to develop relevant prevention strategies, it is necessary to have an understanding of what these harmful effects are.

6.3.3.1. *Definition and evaluation of the health risks associated with atmospheric pollution*

The definition of atmospheric pollution in 1967 focused on its discomfort effects. As time went by, and as people became more aware of the actual impacts of pollution, a better definition of atmospheric pollution could be given. It also became easier to provide a better explanation of the negative effects that pollution had on human health. These negative effects were no longer effects that simply caused discomfort to the population, but were considered as being objective risks from which it was possible to associate certain diseases with the presence of invisible, yet toxic pollutants present in the atmosphere. The damaging effects of these pollutants extend largely beyond human health. With the introduction of LAURE in 1996, it became clear that such pollution with harmful effects could occur not only on a global scale, but also in closed spaces on a local level. It then became a question of

---

9 Definition according to the European Economic Community (EEC) in 1967: There is pollution in the air whenever a foreign substance or a significant variation in the proportion of its components are capable of creating harmful effects (based on current scientific knowledge), or of creating discomfort. Definition according to LAURE 1996: Pollution is the direct or indirect introduction of substances into the atmosphere or into closed spaces by mankind. These substances have damaging consequences on human health, and damage biological resources and ecosystems as well as influencing climate change, deteriorating material goods and leading to excessive olfactory damage.

identifying these risks as being health risks or climate risks [ROU 01a]. This is quite a difficult question to answer because pollution levels fluctuate in relation to changes that occur in the climate, yet pollution itself can influence climate change. Is this a question of increased anthropocentrism or a rediscovery of the importance of the interface that exists between man and his environment? Air quality is symbolic of the changes in the nature of the risks that have been observed in modern society. damages that was once created by natural risks was replaced by pollution that was created by large factories and industries. Nowadays, disasters and the threats of destruction are created by humanity itself [DEL 87]. Health risks are only a small part of the dangers that our modern day society, called "risk society" has had to deal with [BEC 03], a society in which it has become necessary to develop measures to help prevent atmospheric pollution.

Identifying health risks is no longer a question of comparing a pollution value with an air quality standard. Progress made in the fields of epidemiology and toxicology means that it is possible to distinguish between short-term and long-term health risks. An epidemiological study that evaluated the risks of urban pollution on human health was carried out in the region of Paris between 1987 and 1995 [ERP 94]. This study was regularly updated and the results of this study meant that it was possible to understand the short-term impacts (a period of several days) that atmospheric pollution had on human health. These results, in conjunction with numerous studies carried out in the same field, reinforced the idea that there is a clear relationship between air pollution and disease. This relationship is referred to as having no threshold limit, meaning that regardless of what the level of pollution is in the air, there is always the risk of disease. Pollution is dangerous regardless of the quantities of pollution that are present in the atmosphere. This observation questions the importance of certain standards that were set and which suggested that pollution levels that were under a particular threshold did not have a harmful effect on human health. Pollution peaks are no longer indicators that can be used to provide information on health risks. In fact, they are used to indicate the chronic levels of pollution that need to be monitored. The majority of these standards, which are prepared by the CAFE[10], refer to average annual values that indicate the existence of background pollution that needs to be monitored.

6.3.3.2. *The mapping and spatialization of health risks*

The use of GIS means that it is possible to work out the location of people who are exposed to any given level of pollutants in the air. GIS can be used to calculate how much pollution certain members of the population have been exposed to. This calculation is worked out by crossing two pieces of information: first of all by using pollution levels plotted on a map, and second, by using the total number of inhabitants who are resident in the area, which corresponds to a colored area on the

10 *Clean Air For Europe.*

map that represents a particular level of pollution. The result obtained by Airparif[11] shows the number of people who are affected by relatively high levels of pollution. This method for evaluating the risks associated with pollution levels and their effects on the population is quite redundant because it cannot be used to identify the occurrence probability of any disease. Studies that have been carried out have focused on the risks encountered at a local level, whereas in reality, people move from one place to another.

Mapping health risks assumes that a standardized method is used, a method that was developed by the American Academy of Sciences during the 1980s[12]. This method was altered again by the bill that was introduced on February 25, 2005 in the health section of the environmental impact assessment that was carried out and in which an index that measured a population's exposure to pollution was developed. This index uses data taken from the emission inventories, as well as information relating to the spatial distribution of the population, by evaluating any potential exposure to pollution.

This approach calculates the risks of exposure to pollution and to individual pollutants that are attributed with toxicological reference values. There are uncertainties that exist as far as this method is concerned. However, this method can be used to generate useful indications that provide information on exposure levels and which can be used as a basis for further negotiation. However, the information produced, through a very technical language, involves identifying excess risks and identifying probability of diseases occurrence. For example, for a relative risk with a value of 10^{-5}: if an incinerator was located in a nearby area, this would mean that one person out of every 100,000 could develop cancer due to their exposure to the emissions produced by the incinerator. These values are only true for a 70 years exposure long. Although this information is available, the results are only understood by few people belonging to the limited world of people dealing with this subject. This approach needs to be accepted for what it is, an imperfect exercise using data of varying quality to obtain a specific order of magnitude. In terms of management, being able to quantify the risks dramatically changes the nature of the problem. Without any quantified risk, it is not possible to make decisions in relation to rational limit values, and as a result, any decisions that are made will be more severe than necessary pollution [DAB 04].

It is not only human health that is damaged by atmospheric pollution. The phenomenon known as acid rain has highlighted how vulnerable forests are to atmospheric pollution, whilst other studies have shown that agricultural production efficiency decreases in relation with atmospheric pollution, especially during ozone

---

11 http://www.airparif.asso.fr/. An organization that monitors air quality in the region of Paris.

12 National Research Council, Committee on the institutional Means for assessment of Risks to Public Health. Risk Assessment in the Federal Government, Managing the process, Nat. Acad. Press Washington D.C., USA, 1983.

episode. In addition, the way in which vegetation responds to atmospheric pollution provides an indication into what happens when living organic matter is exposed to air pollution. As far as living organic matter is concerned, it's another type of monitoring and method of collecting geographical information.

6.3.3.3. *Bio-indication*

This type of information is associated with the study of the visual damage that is caused by a pollutant on a test plant. It is defined as being the use of responses at all levels of the biological[13] make-up of an organism or set of organisms to predict or to show environmental changes and to monitor their evolution. In other words, it is used to study the [VAN 02].

Using vegetation in this way is based on a recording of macroscopic activity that occurs within an organism so that it becomes possible to examine the effects of pollution. From the studies that were carried out it became clear that the pollutants applied to the test plants led to the development of necrosis (the death of cells, which is characterized by a discoloration of the epidermis), chlorosis (change in color), or other morphological changes, such as change in size or changes to flowering patterns, etc. Many different plants have been used for the purpose of these studies. Tobacco plants were used for the bio-monitoring of ozone[14] (a strongly phytotoxic pollutant). An extremely sensitive variety of the tobacco plant, Bel W3, was used by this method. The presence of ozone on this plant leads to the creation of white-brown foliac necrosis on sheltered tobacco plants that were regularly watered. It is possible to map this type of necrosis, which affects plants and vegetation at a given location, by using a GIS. The GIS will then produce information on the spatial distribution of the harmful effects that are caused by ozone.

In France this type of information gathering is not very common and is not very developed. In France bio-monitoring is used as a complementary approach to the physical and chemical measurements that are recorded. The results produced from bio-monitoring are focused on the effects of pollution exposure on the ecosystem. Due to the ever-changing nature of atmospheric pollution, bio-monitoring offers a range of the latest observation tools that can be used to monitor this phenomenon. Bio-monitoring is an approach that was developed from rigorous scientific research and as a result it can be integrated into evaluation techniques which are used to examine environmental and health risks. In addition to this, bio-monitoring can also be used as a communication tool to provide the general public with information on the pollution of organic living matter.

---

13 But also molecular, biochemistry, cell biology, physiology, tissue biology, morphology, and ecology of the organism(s) in question

14 The use of bio-monitoring is subject to a normalization procedure that is carried out at the French Normalization Association (AFNOR).

The relationship between vegetation and atmospheric pollution can also be applied to the field of agriculture. Crossing pollution levels observed and different types of crops makes it possible to quantify the damage that is caused, and to forecast any loss in agricultural production [CAS 03] as far as the relationship between pollution levels and effects has been pre-established.

The risks associated with pollution change dramatically when the risks are expanded on a global scale. These global scale risks, nevertheless, continue to affect our daily lives. Geographical information is particularly able to acknowledge these changes from the local level to the global level.

6.3.3.4. *Greenhouse gases and the global risk*

The life span of certain pollutants and their build-up (accumulation) in the lower layers of the atmosphere over a long period of time have completely changed risk-management strategies, which are developed to prevent the risks associated with pollution from developing. Prevention is no longer a question of introducing an avoidance policy to pollution plumes, it is now about limiting the emissions that are produced at the source. Over the past few years industrial waste has dramatically decreased, and as a result of the decentralization of certain administrative tasks in France, it is now up to the individual regions in France to limit the amount of greenhouse gas emissions. In addition, it is also up to the general public to change their habits in order to save energy. However, the inertia associated with the build-up of pollutants in the Earth's atmosphere makes climate change inevitable. In order to save more energy a climate adaptation strategy needs to be introduced and this can be in the form of energy saving houses, growing crops locally, etc.

Risk management and examining the risks associated with pollution form part of a science that is still in its early stages. Trying to prevent such risks from arising is linked to the paradigms that have been developed from improved modern technology. Knowledge of the different impacts that pollution can have on the environment and on human health is required so that prevention strategies and climate adaptation strategies can be developed. However, data providing information on the different impacts that pollution can have on the environment and on human health is much more complex than the data used to provide information on pollution levels, especially as far as traditional pollutants are concerned. Publishing information on pollution measurements with no explanation of their consequences may be harmful and useless to (stigmatization) the population that is exposed to such pollutants. Furthermore, the information produced is regarded as being unhelpful and unnecessary. The information should relate to the following different paradigms: prevention, alert, monitoring, and adaptation. From the study of the different sources of pollution to the study of emissions and impacts of pollutants on the environment and human health, an entire chain of models is created and generated by information that is produced by multidisciplinary scientific data. The tools developed within this chain uncertainty grows during the demarche goes on. However building action takes place in this context of scientific uncertainty.

6.4. The geography of science and action

Before GIS is implemented on a large scale it should be noted that its use is rather limited, irrespective of the method used to gather data on air quality. This information can then be integrated within the GISs. These limitations associated with the study of pollution and climate should be used as a basis from which constructive decisions can be made. This decision, as consensual as possible, has to be open to different points of view.

6.4.1. *Limits associated with such tools*

The accuracy of documents produced by GIS, often improved by color areas that can be seen on maps, should not hide the fact that certain restrictions can be applied. Pollution levels, which are represented on a map, correspond to information that has been gathered for a particular area. This is why the transition from one pollution level to another is represented on a map by a change in shade or color. These tools have large uncertainties, especially when they are used in the field of spatialization.

6.4.1.1. *A GIS freezes atmospheric activity in time*

A GIS will record a particular event that is taking place in the atmosphere regardless of its duration. As far as air quality is concerned, there can be more temporal variations associated with rapid evolutions of time than there are spatial variations associated with emissions of pollution and topographic conditions that influence dispersion of the pollutants. A GIS measuring air quality may depict different situations depending on the type of weather conditions in which it is used. For example, average, harsh or unstable weather conditions lead to the presence of a different mix of pollutants in the atmosphere. The maps are a snapshot of chosen weather patterns for which average weather conditions do not exist.

6.4.1.2. *The regional identification of the sources of atmospheric pollution*

The pollution levels recorded take into consideration different phenomena that take place on many different scales (see Figure 6.4). Local pollution levels contribute to a small part of these recordings because it depends on background pollution whose sources are far from the measuring site, and the data provided from these sources are diverse and not recent. This fact discourages lots of specialists working in the field, because despite all their hard work geared towards reducing pollution levels, there is no actual spectacular decrease in the levels of pollution.

The lack of results produced by this method does not match the amount of time and investment that is required and that has been made in trying to reduce pollution in the following areas: public transport, heating systems, and industrial processes, etc.

However, modeling atmospheric pollution on a small scale leads to the creation of spatial discontinuities that indicate that there may be hot spots present. The presence of hot spots means that the issue of atmospheric pollution is far from being resolved, despite the fact that there has been a general decrease in the presence of certain pollutants in the atmosphere. Although background pollution tends to be uniform over a large area, hot spots tend to be more localized and can exist next to factories or main roads. Certain maps show these spatial variations, from these maps it is possible to identify the more polluted areas, and it is also possible to see that people living in these zones are exposed more to risk pollution.

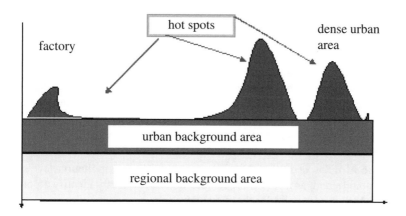

Figure 6.4. *The regional identification of the sources of atmospheric pollution (see color section)*

It is not enough to produce a simple report on the objectification and identification of inequalities that exist as far as atmospheric pollution is concerned. It is also necessary to provide evidence that remediation elements exist. It is not possible to identify spatial variations in air quality without having any additional information available. Atmospheric pollution is defined by the impacts that it has on people, vegetation, and on the planet as a whole. In order to overcome these impacts it is necessary to understand what they are and why they occur. How can tools that have been developed to help provide a better understanding of, and overcome the risks associated with pollution, be seen as an aid? The world of geography, which is also known as interface science, plays an important role in trying to contribute to this knowledge provide a solution to this question.

6.4.1.3. *Difficulties associated with the mapping of health risks*

Mapping health risks depends on the exact amount of pollution that a population has been exposed to. The idea of exposure to pollution, especially as far as maps and GIS are concerned raises several questions, and in particular, questions relating to the idea of mobility, because mobility is a characteristic associated with urban populations, whilst maps provide information about the area in which people live. Exposure levels vary greatly depending on different reasons including: the mobility of the urban population, the conditions of their homes, and interior pollution that measurements used to record surrounding pollution do not take into consideration.

– It is extremely difficult to carry out research on an air sample and to model levels of exposure to pollution. Research in this field combines both fluidity and environmental dynamics. There can be a big difference between the actual population living in a region and the proportion of this population that has been exposed to certain pollutants. This can be explained by commuting and population travel from home to the office. The actual amount of pollution that a population is exposed to may vary substantially and this can be explained by the amount of time the people spend in the area, and by the location or other characteristics associated with the buildings in which they find themselves when they are in the area. The conditions of a building, as well as the pollution that exists inside the building, can strongly influence the amount of pollution that people are exposed to. Studies into the summer heat wave that occurred in Paris in 2003 have shown how the conditions of the buildings in the city influenced the death rate of those who died due to the heat wave [INV 04]. These studies highlighted the social and vertical stratification that existed in 19th century cities in France, and which could only be recognized by 3D (3D) GIS.

– The use of maps or GIS makes it possible to compare pollution levels with normal air samples and, as a result, it becomes possible to partly identify a potential health risk, which is only valid for the surrounding air.

– Epidemiological studies carried out to study the health risks associated with exposure to pollution use values that are measured by the AASQA, meaning that there is no need for concern about the actual amounts of pollution that are inhaled by the inhabitants of a city. It is safe to assume that the health risk associated with air pollution is low; however, large study samples are required in order to produce reliable results. Measurements recorded in the surrounding air are not enough on their own to define a health risk. These measurements only give an approximate indication of the exact amount of exposure that an individual or group of individuals has had to atmospheric pollution.

For any type of disease to develop, the pollutants must penetrate an organism. The effects of pollution become much worse as the actual dosage[15] of pollution that a person has been exposed to increases. Exposure to pollution corresponds to the exact dosage of pollution that a person has been subjected to and can be measured according to the following three factors:

– the concentration of pollutants present in the atmosphere: car drivers, cyclists, traffic users, and children (who find themselves in a location that is close to car exhaust pipes) are exposed to a higher dosage of pollution. In certain streets that are protected from the wind, pollution levels may be higher;

– the length of time a person is exposed to pollution: city dwellers normally spend more than 80% of their time in buildings or in transport. However, it is the people who work outside who are exposed the most to pollution;

---

15 Quantity of a pollutant inhaled by a person.

– physical activity: physical effort is accompanied by an increase in pulmonary ventilation and because of this more pollutants are able to enter a person's respiratory system.

Data are available that provide some information on the first of the three abovementioned factors. This can be explained by the fact that the measurements recorded do not take the quality of air inside buildings into consideration, even if city dwellers spend, on average, more than 80% of their time at home or in the work place.

The geographical information produced nowadays is no longer subject to the peremptory and authoritarian affirmations that were once made. It is now produced in such a way that many different viewpoints can be considered, and can be used for the creation of a common model, in other words it is now used as the basis of governance.

6.4.2. The difficult task of reducing pollution levels. Can information be used as a governance tool?

The issue of atmospheric pollution can be used as a great example to highlight the ideas and issues that are associated with governance. Brodhag defines governance as: In the context of sustainable development governance is considered as being a process of collective decision-making which does not involve only one authoritative entity. In a complex and uncertain system, for which the different issues are linked together, none of the people involved in the system have all of the necessary information, nor do they have the authority to properly manage a long-term strategy. This strategy can only be created as the result of co-operation between the institutions and the different parties who are interested in developing the strategy. In such a co-operation each entity is entitled to exercise their full responsibilities and execute their power[16].

Knowledge and the distribution of information to a wider public suggests that more people are becoming involved in the decision-making process and this complicates actions that need to be carried out in a domain in which everything is very unclear and in which many different spatial-temporal scales overlap. The diversity of those members of the public who receive information about pollution levels implies that there has been a certain amount of standardization of the measurements, and of the data. In addition to this, demanding protocols have been introduced to which the metrologist has to adhere.

---

16 Translation of the French definition given by Christian BRODHAG, Florent BREUIL, Natacha GONDRAN, François OSSAMA, *Dictionnaire du développement durable AFNOR*, 2004 edition.

6.4.2.1. *An increase in the number of people involved*

The progressive emergence of the issue of atmospheric pollution in the political world has meant that there has been an increase in the development of a certain number of institutions, from the AASQA through to global conferences on climate. However, air quality affects each and every one of us as far as environmental health is concerned. Atmospheric pollution forms a large part of the French government's National Plan on Environmental Health (PSNE), which was adopted in 2004. Making the general public aware of the quality of the air that surrounds them, as well as getting them involved in fighting against air pollution are the main objectives of governance.

– *Europe plays a key role*: ever since the signature of the Single European Act in 1986, the environment has become one of the major strong points of the European Community. The European environmental policy is mainly based on the use of regulated equipment. This policy also relies on the use of a long-term strategy, as is the case for the program known as CAFE[17]. CAFE is made up of numerous different sections whose economic benefits have been carefully evaluated.

– The introduction of the European Community's Environmental Policy has made it possible to encourage the creation of an international strategy: the Vienna Convention (March 22, 1985) and the Montreal Protocol (1988) on aerosols were precaution strategies that were drawn up to reduce emissions of pollutants, and on March 14, 1990 a climate adaptation strategy was developed in order to adapt to the consequences that the strategies introduced in 1985 and 1988 had on the environment. At the beginning of the 1990s Europe became a continent to be reckoned with as far as global pollution and the greenhouse effect were concerned.

– The law known as LAURE states that issues relating to air quality are dealt with on a national level, even if the regional authorities deal with these issues on a regional level: the State and its public establishments, the local authorities and private individuals are responsible for adhering to the environmental policy whose aim is to make sure that everyone has the right to breathe air that does not damage human health[18]. The result of a debate on hygiene measures, held by a certain number of important French town councilors, saw a rapid introduction of air quality monitoring in France with the development of a technical network of measurement sites that was used to measure how clean the air was with the aim of meeting the standards set by the European Commission. During this time, air quality monitoring was no longer supervised by the Ministry of Health, and this power was transferred to the Ministry of the Environment. Some problems were encountered because of this shift in power, but it was seen as a defining moment for French politics and culture. It was the State that developed technical and scientific rationality in France, with the help of the DRIRE. On a European level, the French government is responsible of the measurements viability that have been recorded in France and is responsible for respecting the European standards that were introduced. France is

---

17 *Clean Air For Europe*.
18 Article 3 from LAURE.

also proud of being able to participate in talks in which these standards are set. Nevertheless, with the return of epidemiology and large-scale studies, such as ERPURS [ERP 94][19], health risks have once again become the key objective of air quality monitoring [JOU 07].

Monitoring networks: the networks used to monitor air quality have been reorganized since the introduction of LAURE and recognized as the place around which air quality is focused. Within the four different boards that make up the Board of Directors, there has also been some controversy as far as the fight against pollution is concerned. Such controversy has included the debate on decentralization, which focuses on the distribution of power from the State to the local authorities, and the debate on the monitoring air quality role, between knowledge and action with the risk to mixing expertise and management [LAM 07].

The ambiguity of the role of the AASQA is due to the difficult relationship that exists between communication and action. The ASSQA plays an important role as far as information is concerned, but were they created to be used in governance place? The recent report written by Senator Richert also raises this question [RIC 07].

6.4.2.2. *Geographical information: a tool for communication or for action? is it used to actively fight against rising pollution levels?*

The AASQA is the organization that possesses and produces most of the information that is available on air quality. How can such an organization produce and hold such a tool box without changing its focus to concentrate more on pollution management strategies? Improvements in the knowledge that experts in the field of climatology have [KOU 00] highlight the weak distinction that exists between the two areas of expertise and management. There are no real boundaries between these two areas, which brings into question the process of monitoring.

Maps showing the levels of pollution that have been recorded for a particular area lead to problems in a political context. This is because the maps are associated with a process of power and are used to monitor, control, and guide individual members of the population living in areas prone to high pollution levels as to what they can do in order to reduce pollution levels. If these urban rules are not adhered to, the maps can then be used as a method to punish those people who do not respect the urban rules that are set [FOU 75]. The maps are also used to integrate the notion of risk into planning strategies that focus on air quality[20]. Are these types of maps now a product of climatologists' knowledge or are they used as a management tool?

---

19 Urban pollution.

20 This notion of the power that maps posses can be applied to plans that were introduced by the French government to reduce the risks of flooding, through the introduction of the risk prevention plan (PPR) and the risk prevention plan for flooding (PPRI).

The recent use of GIS is likely to dramatically change the way in which atmospheric pollution is managed. This is especially the case for the production and distribution of scientific documents that identify areas in which there are health problems. The forecast of a particular event which is due to take place, once people are aware of it and as soon as it has been advertised, can only modify the event in question [DUP 02]. As soon as it is developed, the distribution of such scientific documents can be interpreted in two ways by two different groups of people; first by the people who developed the document, and second, by the readers of the document. For the people who created the document, the document is seen as being perfect and the data introduced in the document are gathered by recording measurements and creating models that are displayed in graphs, but only after the data have been rigorously analyzed. As far as the readers are concerned, the document is seen as a finished product, an image whose scientific supports have been removed [DAG 03]. The model used in the end is a simplification of what really occurs. These hypotheses should be used as a grid used to analyze and evaluate the finished documents. The characteristics of these maps and the peremptory characteristics of the color areas on the map make it possible to forget about slight differences in shading when the results are being interpreted. However, a large part of the influence that these maps comes from the fact that they are used in a branch of science that is seen as being neutral, conforming to social demands, yet at the same time it also refuses to conform to these demands [HAR 95].

6.4.2.3. *The map, a communication tool or ambiguity created by the idea of monitoring?*

Analyzing the maps which are created [BOU 02] shows the ambiguity between the maps that are created (and which provide information on pollution levels in a particular area), and the actual scientific meaning of the pollution levels that are recorded and which need to be taken into consideration, as well as the effect this information will have on the general public when it is made available to them. The quality of the maps which are produced or made available on the internet should not be affected by the quality of color printing or by the care and effort that has gone into creating the graphics. However, the maps created are no longer focused on reliability, nor are they focused on the precautions that must to be taken as far as their own interpretation is concerned. The boundaries that exist between different color areas on the maps are smooth, because in reality the transition between two different levels of pollution is not differentiated by a sudden change in the amount of pollution present. This smoothing is due to a political precaution which avoids seeing the exact location of where people live.

The fact that certain maps are published also shows the limits associated with the transparency of the information found in the maps. The results produced by models that measure the impact that the plume produced by different factories has on the ground can only be communicated to the inhabitants of the area if the results are accompanied with commentary on the different hypotheses that have been developed regarding the negative effects of pollution on human health. In addition, an evaluation of the potential risks incurred also needs to be published and then communicated to the inhabitants of an area along with the results of the modeling process. However, the use of these maps does present one main disadvantage, and that is the fact that people who live in disadvantaged ecological areas tend to be stigmatized.

The idea of monitoring air quality is ambiguous as it plays a key role in both the scientific and political worlds. For example, the organization known as the AASQA possesses both a scientific and political side. As far as science is concerned, the AASQA can produce lots of documentation, such as maps. However, as soon as the maps are produced, it is possible to see the political side of this organization in action as it must ensure that all of the documents have been received by the people for whom they were created. The aim of the documents is to show the knowledge and the competences that the monitoring associations possess, rather than simply identifying and locating problem areas.

The problem with the current system is that the documentation is focused too much on the knowledge that specialists in the field have, whilst there is a great need to focus more on the action of preventing air pollution. The domain of monitoring air quality is not limited to technical know-how, because as far as the environment is concerned, the word knowledge means action. In this domain it is necessary to be pro-active rather than re-active.

How can these factors, which are very important as far as expertise in the field is concerned, be used more actively in the fight against pollution? This question is very ambiguous and has raised several issues as illustrated by the work carried out by AIRPARIF[21] on the heat exchanger located at Bagnolet, Paris, France[22]. This study has shown that for certain pollutants, pollution levels are notoriously higher the closer one gets to the exchanger or the closer one gets to the suburb's periphery in comparison to the levels of background pollution that are present. The size of the area that was studied corresponds to a population of approximately 70,000 people. The data collected in relation to how emissions affect health, provide no explicit information on how such emissions affect health. To what extent does such a heat exchanger affect health? In the event of an extreme risk, the knowledge and know-how that specialists possess is completely counterproductive (except if there are any attempts at remediation). It is said to be counterproductive because in the event of an extremely high risk, there is the fear of stigmatizing a particular region. In other words, the region will be seen as being a disadvantaged region as far as the following factors are concerned: social, economic, health, or environmental factors. Such negative publicity can lead to the creation of major inequalities, but it should be borne in mind that that pollution levels of the surrounding air is not the lone factor responsible of such a situation. The exact amount of exposure to pollution

---

21 http://www.airparif.asso.fr/page.php?rubrique=projets&article=echangeur.

22 Bagnolet is a suburb found in the eastern side of Paris. Many different examples could be used here. The issues arising from the existence of such a heat exchanger are the same issues that arise in the surrounding areas of factories.

that people living in a particular area are faced with needs to be modified. This is possible by taking into consideration the actual levels of pollution that exist inside buildings, because Parisians spend more than 80% of their time inside offices when they are not at home[23]. Preventing such pollution from taking place appears to be a difficult task. It is not a question of moving those people who might be exposed to pollution to another area, and who have not been identified as potentially being affected by pollution. Neither is it a question of removing the heat exchanger. This leads to the following difficult question: is it possible to have a local strategy that can be used to improve pollution levels (for example, by focusing working on the periphery of the city)? The corresponding map produced not only raises scientific doubts, but it also begins to raise doubts regarding preventing pollution.

The same types of questions are also raised for those areas that surround factories. For example, maps that show the concentration of benzene levels in the air have led to the creation of a GIS at Drocourt, in the French region of Nord-Pas-de-Calais. The measurements, which were taken several times by this GIS, have shown that there is an abnormally high level of benzene present in the air. Two industries that were suspected of emitting benzene were located in the areas under investigation. Furthermore, tests carried out to see whether there were any health risks associated with the emission of benzene showed that there was a clear increase in the amount of benzene produced in the area. As a result, one of the factories closed and benzene levels became significantly lower. The documents that were produced and provided an analysis of the investigation showed that there were two main problems: first, the only solution to improve sustainable development might not have been to close down work places and thus increase unemployment, and second, there was the issue of historical pollution. In what state is the health of members of the public who spent a large part of their lives exposed to high levels of pollution? Even if the factory has closed down, some members of the public have nonetheless spent years of their lives exposed to such pollution. This pollution, which has accumulated over time, still remains in their bodies.

Mapping pollution hotspots raises another problem: identifying the different sources of pollution. This problem is made up of a different number of collective processes which are associated with atmospheric pollution. This problem is also linked to issues associated with scale effects which make the individual analysis of different intervention levels difficult to carry out. Progress has been made in a branch of modeling known as inverse modeling, and the progress made in this field makes it easier to trace the source of a particular pollutant that has been emitted.

Nevertheless, expensive measures adopted to reduce pollution levels on a local scale have not been very productive, which is disappointing. AIRPARIF examined the policy on public transport introduced by Paris City Council, and found that

23 The influence that the characteristics of buildings has on interior pollution was highlighted by the Institute of Health Monitoring (INVS) by studying the distribution of dead bodies that were found as a result of the summer heat wave in 2003.

getting everyone to use public transport instead of driving did not have a huge impact on air quality; there was only a very slight decrease in pollution levels. It is possible that the pollution, which only existed in the center of Paris, has started to extend out to the city's peripheries. Much research still requires to be carried out on this theory.

The use of a GIS to create a prevention plan (formed on the basis of the documents produced, which provide information on the amount of pollution that is present in the atmosphere) requires the collaboration of many different people who are involved in the entire process. These people range from the model designers right through to the people in charge of health and administration, as well as the politicians and the inhabitants of the areas in question. It is possible for a GIS to become a governance tool by using all of the different people who are involved in the fight against pollution.

6.4.3. *Is geographical information used as a governance tool?*

The introduction of governance has led to an increased number of technicians leaving the world of teaching to take on more support and intermediary roles[24]. The large number of communication services that exist in towns means that for the moment elected politicians are able to create positions for those technicians who are changing vocation. Thanks to this change in vocation it has become possible for technicians to communicate information to the general public in simple terms. Trying to decode extremely technical information and make it available to the general public as well as to decision-makers in a language that can be easily understood is a real challenge. Complicated, technical language should be avoided in order to create a real sense of governance between the different groups of people who are affected by the quality of the environment that surrounds them, and thus by the quality of the surrounding air. There are occasions when the technicians who work in this field do not understand each other, meaning that the language that they use will not be understood by the general public. Governance is often hindered by the technicians and their backgrounds as they are incapable of producing information that is aimed at the general public. Each individual inhabitant of a town or city needs to be involved with governance, so that this method can be applied to the problems associated with air pollution. On one hand, governance needs to feed public debate on a particular topic, and on the other, the general public needs to become more involved in governance. This idea of governance stems from the world of politics, and should incorporate technical knowledge that can be used to overcome the problems.

After being used as part of a long trial period, would it be possible to integrate mass communication methods (made up of GIS) into more general methods that are used to improve the quality of the environment on all levels? What the inhabitants of towns and cities want to know about pollution levels and what they can do to

---

24 State plan adopted in the region of Nord-Pas-de-Calais.

reduce them is still somewhat unclear as their demands are seen as being rather contradictory. For example, if car pollution in a given town is a major problem, then why are none of the inhabitants of this town capable of leaving their cars at home? This issue highlights the public's inability of being able to live without a car, and also shows the limits of power that the local authorities possess. The decrease in the gap between knowledge and politics is a major issue which can be resolved by the directors of pollution management networks, since it is these directors who have access to the tools which can be used to evaluate the methods used for pollution prevention. The directors also need to work on removing the collusion that exists between the media and the technical systems that are in place to monitor air quality; they also need to replace information with speech and to restore pride in political policies that are introduced, in both the scientific and technical works, and whose main aim is to reduce pollution levels. Finally, the directors of the pollution management networks also need to introduce a new type of democracy into today's advanced technological societies in order to respond to the issues that these societies are faced with[25].

If new public policies are created, this will mean that politicians will have to make more difficult decisions. Different debates need to be held so that the people can work together in order to decrease the toxicity levels of the invisible pollution that exists in the atmosphere. The views and opinions of the people involved in such debates will not always be unanimous, and some people may even be seen as being unpopular, especially when it comes to responding to different criteria to improve public health. The information available and produced from different studies that have been carried out cannot be used as an excuse to explain why there been no decrease in the levels of pollution that are present in the air. The government is going to have to take control of what is happening as far as the interests of the general public are concerned [ROU 04].

6.5. Conclusion

Monitoring atmospheric pollution can be considered as being symbolic of techno-science that has been supported by the State. The alliance that exists between the two entities (the State and the world of science and technology) is typical of a strongly centralized country that benefits from the support of large state bodies, such as the groups of state-funded engineers who also work as directors [CAL 05] and who work to develop powerful tools that can be used to monitor atmospheric pollution. Combining these two entities allows different parties to work together; on one hand, there are the agents and technicians who work for the State, and on the other, there is the general public[26].

---

25 P. Roqueplo, Critical and Perspective Thinking in Ruptures créatrices, section Trends, *Les Echos*, p. 567-597, 1999.
26 COANUS T., PÉROUSE J.F., *Villes et risques*, Economica Anthropos, p.27.

Due to the large number of concerns that the general public has regarding environmental risks, the general public demands that more information be broadcast to them, on a more regular basis. How relevant is this technically-sounding information which is published and then distributed to the general public, especially when it does not include any of the elements used to examine the different hypotheses that exist on pollution management? GISs raise the issue of the regional management of the complex non-spatial system known as air quality. The fact that it is impossible to spatialize measurements that show the amount of pollution an individual has been exposed to questions the relevance of the spatial tools that are used. Would it not be more useful to have tools that highlight other fields rather than geographical areas that are affected by different types of pollution? For example, the tools could provide information on the amount of pollution farmers are exposed to and how this affects their health, whenever they are exposed to different levels of pesticides.

The field of atmospheric pollution is continually evolving. Information published by the AASQA needs to be made more complete by adding information that relates to the quality of air inside buildings and to the monitoring of greenhouse gases. As Richert's report suggests [RIC 07] the information available needs to be able to evolve over time. Should different factors that are associated with air quality be included in this information? Is the LAURE law, which was introduced in 1996, changing and evolving into smaller atmospheric environments?

The increased focus on the individual, as is the case with measuring the amount of pollution an individual is exposed to, increases the uncertainty that exists as far as spatial pollution measurements are concerned [CHA 07]. Is it not the case nowadays that we are experiencing a revival of environmental health policies introduced to protect the individual from the harmful pollutants present in the air? This notion of protecting the individual involves a greater focus on the toxicity of the polluting products rather than focusing on pollution emissions, which are under control in developed countries. People who move from one area to another will also become increasingly responsible for their own health, rather than their health being the responsibility of a particular region.

6.6. Bibliography

[BEC 03] BECK U. *La Société du risque - Sur la voie d'une autre modernité*, Flammarion, Champs, 2003.

[BOU 02] BOUTARIC F., LASCOUMES P., RUMPALA Y., VAZEILLES I., *L'obligation d'information, instrument d'action publique. Surveillance et délibération dans la lutte contre la pollution atmosphérique*, Paris, CEVIPOF, 2002.

[CAL 05] CALLON M., LASCOUMES P., BARTHE Y., "Les citoyens ordinaires: une menace?", *Problèmes politiques et sociaux*, no. 912, 2005.

[CAS 03] CASTELL J.F., LEBARD S., "Impacts potentiels de la pollution par l'ozone sur le rendement du blé en Ile-de-France: analyse de la variabilité spatio-temporelle", *Pollut. Atmos.*, vol. 179, pp. 405-418, 2003.

[CHA 07] CHARLES L., La pollution atmosphérique, entre individu et collectif: mise en perspective sociologique), in CHARLES L., EBNER P., ROUSSEL I., WEILL A., *Evaluation et perception de l'exposition à la pollution atmosphérique*, Paris, La Documentation Française, pp. 121-143, 2007.

[DAB 01] DAB W., ROUSSEL I., *L'air et la ville*, Paris, Hachette, 2001.

[DAB 04] DAB W., "L'étude d'impact sanitaire: un outil de gestion des risques sanitaires liés à l'environnement", *Ann. Mines*, pp. 57-59, 2004.

[DAG 03] DAGORNE A. et DARS R., *Les risques naturels*, Que-sais-je?, PUF, Paris, 2003.

[DEL 87] DELUMEAU J., LESQUIN Y., *Les malheurs des temps: histoire des fléaux et des calamités en France*, Larousse, 1987.

[DUP 02] DUPUY J.P., *Avions nous oublié le mal?*, Bayard, 2002.

[ERP 94] ERPURS, *Impact de la pollution atmosphérique urbaine sur la santé en Ile-de-France*, Paris, Regional observatory for health for the region of Paris, 1994.

[FES 97] FESTY B., "Fixation des normes de pollution de l'air", *Pollut. Atmos.*, 1997.

[FOU 75] FOUCAULT M., *Surveiller et punir. Naissance de la prison*, Gallimard, Paris, 1975.

[HAR 95] HARLEY B., "Déconstruire la carte", in GOULD P. and BAILLY A. (eds.), *Le pouvoir des cartes, Brian Harley et la cartographie,* Anthropos, 1995.

[INV 04] INVS., *Vague de chaleur de l'été 2003: relations entre température, pollution atmosphérique et mortalité dans neuf villes françaises*, study report, Institut de veille sanitaire (*InVS*), September 2004.

[JOU 07] JOUAN M., "La pollution atmosphérique: un enjeu de santé publique, quelles actions?" in CHARLES L., EBNER P., ROUSSEL I., WEILL A., *Evaluation et perception de l'exposition à la pollution atmosphérique*, Paris, La Documentation Française, pp. 109-115, 2007.

[KOU 00] KOURILSKY P., VINET G., *Le principe de précaution*, Paris, Odile Jacob/La Documentation Française, 2000.

[LAG 99] LAGRANGE X., GODLEWSKI P., TABBANE S., *Réseaux GSM-DCS*, 4th edition, Hermès, 1999.

[LAM 07] LAMELOISE P., "Quelle surveillance, pour quel public?" In CHARLES L., EBNER P., ROUSSEL I., WEILL A., *Evaluation et perception de l'exposition à la pollution atmosphérique*, Paris, La Documentation française, pp. 21-29, 2007.

[MAR 04] MARTINET Y., "Conception, validation et exploitation d'un cadastre d'émissions des polluants atmosphériques sur la région Nord-Pas-de-Calais", PhD thesis, French College of Engineering, Douai , 2004.

[MIE 00] MIETLICKI F., GOMBERT D., "Alerte du 30 septembre 1997 et première mise en place de la circulation alternée: première évaluation faite à l'aide de l'outil de modélisation SIMPAR", *Pollut. Atmos.*, vol. 165, pp. 69-84, 2000.

[NOL 00] NOLLET V., SCHADKOWSKI C., HUE S., FLANDRIN Y., DECHAUX J.C., "Elaboration d'un cadastre d'émissions de polluants primaires dans la région Nord-Pas-de-Calais", *Pollut. Atmos.* vol. 165, pp. 109-119, 2000.

[RAM 04] RAMBAUD J.M., "Les effets sur la santé de la pollution atmosphérique induite par les industries et les infrastructures: procédures d'évaluation et débat public", in *Joint Conference APPA-ADEME*, Pollutec Exhibition, Lyon, 2004.

[RIC 07] RICHERT P., *Qualité de l'air et changement climatique: un même défi, une même urgence. Une nouvelle gouvernance de l'atmosphère, Rapport de mission parlementaire*, Paris, La Documentation Française, 2007.

[ROU 01a] ROUSSEL I., "La difficile mais nécessaire territorialisation de la qualité de l'air", *Pollut. Atmos.*, vol. 169, pp. 75-85, 2001.

[ROU 01b] ROUSSEL I., "La pollution atmosphérique, risque climatique ou risque sanitaire?" *The International Association for Climatology Publications*, vol.13, pp. 538-547, 2001.

[ROU 04] ROUSSEL I., CHARLES L., "Peut-on parler d'une gouvernance de la qualité de l'air?" in Scarwell H.-J., Frenchomme M. (coord.), *Contraintes environnementales et gouvernance des territoires,* La Tour d'Aigues, Aube editions, 2004.

[TRE 06] LE TREUT H., "Le diagnostic scientifique de l'alerte à l'éclairage des choix", *Ecol. Polit.*, vol. 33, pp. 21-37, 2006.

[VAN 02] VAN HALLWYN C., GARREC J-P., *Biosurveillance végétale de la qualité de l'air*, Tech & Doc, 2002.

[VLA 99] VLASSOPOULOU C., "La lutte contre la pollution atmosphérique urbaine en France et en Grèce. Définition des problèmes publics et changement de politique", PhD thesis, Pantheon-Assas Paris II University, 1999.

Chapter 7

Geographical Information and Climatology for Hydrology

7.1. Hydrological problems of today's society

Throughout history different societies have had to deal with problems associated with water, by either protecting themselves from it or by trying to benefit from its advantages. Ancient Chinese, Egyptian, and Indian societies had to carry out major hydraulic work in order to protect themselves from floods. The work that these societies carried out also enabled them to have access to water as well as being able to provide water for their herds and crops. The Romans left traces of their vast knowledge as far as the world of hydraulics is concerned and evidence of this can be seen by visiting their ancient sites. It is clear that throughout the course of history there have been different developments in the field of hydraulics. Unlike today, however, the hydrologists of the past most certainly worked on a trial and error, or observation basis, and there was certainly no research carried out before a new development was created (for example, nilometers were used to measure the water levels of the river Nile in high Egypt several centuries before the birth of Jesus Christ). However, empiricism and common sense were not always used [HUB 84].

It was not until the modern era and the 20th century, in particular, when proper modern rules stemming from the worlds of physics (hydraulics) and statistics (hydrology) were used. The term hydraulics may be well known, but the term "hydrology" is not so well understood. Hydrology is the study of the Earth's water cycle. The term surface hydrology is used for the superficial part of the water cycle. For the purpose of this study it is surface hydrology that we are interested in. The aim of hydrology is to examine the quantity of water flowing on the Earth's surface. If this quantity of water is known, the field of hydraulics can be used to work out water levels, and the velocity relative to the environment in which the water is found [GOU 05].

Chapter written by Jean-Pierre LABORDE.

7.1.1. *Forecasting and predetermination*

Depending on the aims of studies carried out in this field, a hydrologist can be faced with many different problems associated with the forecasting and predetermination of water levels. More often than not, hydrologic analyses are carried out with the aim of creating hydraulic structures. The analyses are therefore carried out and formulated in such a way that their results can predict the probability of failure of the hydraulic structures. What happens if we want to create a rainwater drainage network that overflows, on average, 10% of the time? The aim of the hydrologist here is to determine decennial flood levels. There is one chance in 10 that these levels will be exceeded in any year. As far as the protection of a dam is concerned, the risk of it exceeding its limits 10% of the time is much too high due to the disastrous effects this would have on human life. This figure of 10% therefore needs to be reduced, and in order to do this, it is necessary to study the flood levels in the area for the past 1,000 to 10,000 years [GUI 67].

Regarding water supply, it is often necessary to work on improving the structures that are already in place (for example, straightening and strengthening the dams for storing water) [ALE 96]. Drought or rationing irrigation are then translated into economic terms to be balanced with necessary investments. Several comparative studies have been carried out in this field and we ask the hydrologist volume adjustments for a wide range of failure rates. This type of problem is part of the category known as predetermination in which a hydrological phenomenon (such as flood, annual water supply, etc) is associated with a probability level.

The second problem that occurs is the problem of forecasting. Problems associated with forecasting tend to occur less than the problems associated with predetermination. As far as forecasting is concerned a particular risk is not associated with a particular phenomenon, but a particular phenomenon is associated with a particular date: for example, what will the water level of the River Seine in Paris be tomorrow? The problem associated with forecasting is a management problem that cannot be applied when hydraulic structures are being developed. However, statistics play an important role as far as forecasting is concerned because it is necessary for a forecast to be associated with a confidence index. The answer to the question, what will the water level of the River Seine in Paris be tomorrow, is more than likely to be: 950 m^3/s and there will be a 70% chance that the water level will be between 880 and 1,100 m^3/s.

The aim of this short introduction is to make the reader aware of the different problems that a hydrologist has to deal with, and to make the reader aware of the fact that these problems are linked to statistics that provide information on a long series of water level observations. Such information is not very widely available and this justifies the fact that the field of hydrology is becoming increasingly closer to the field of climatology in which temporal distance is much greater.

7.1.2. *Water flows at the outlet of a drainage basin*

We have seen that the work of a hydrologist involves examining flows for a particular area for which improvements need to be made. These improvements are not made to any area, but rather to sites in which water flows are concentrated. In other words, the improvements are made to sites that are part of the hydrographic network. Such sites may be permanent or temporary, and are more than likely to include points located in the outlet of a drainage basin. A drainage basin is defined as being the area in which water discharge is transferred to the basin's outlet.

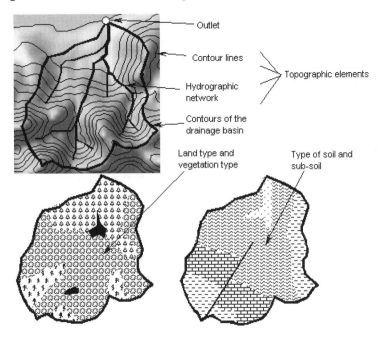

Figure 7.1. *The drainage basin and its components*

The drainage basins included in this study range in size from a few hectares (urban hydrology) up to several million square kilometers. The data available that provide information on flows are sometimes not enough for a statistical study to be carried out; therefore, the hydrologist uses climatological data because this type of data provides a larger temporal distance. Therefore, the drainage basin acts as a filter between climatological data and the transformation of this data which can be used to measure water flows.

The different elements of the drainage basin, as well as the geographical information used to generate these different elements, are maintained by the use of Digital Elevation Models (DEM) for topographic elements, and by the use of Geographical Information Systems (GISs) for other elements, such as land-use type, vegetation type, geology, and pedology.

7.1.3. *Global or distributed approaches*

Converting climatological data into hydrological data is a difficult process due to the spatial variability of the climatological phenomena that the hydrologist would like to use, such as rainfall and temperature. Not only is it made difficult because of the aforementioned spatial variability, but also because of the different environments that the data are recorded in (such as the nature of soil and sub-soil, relief etc), and because of the hydrologists lack of understanding of the laws of physics that govern the relationships between rainfall, evaporation, and water flows, etc.

If we take the example of a relatively small drainage basin (with an area of several hectares to a few hundred square kilometers) the following assumptions can be made: precipitation and evaporation are uniform for the entire drainage basin. Under these conditions, the drainage basin is considered a system whose input is seen as being unique chronicles of data that relate to different climatic factors [LAV 97].

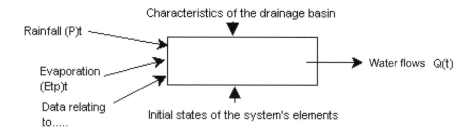

Figure 7.2. *A diagram representing a global model*

However, it is not possible to use the global approach for large drainage basins where rainfall is no longer considered as being uniform for the entirety of the basin.

The drainage basin is divided into sub-basins that are small enough so that, individually, they can be represented by a global model. Water levels associated with each of these sub-basins are then combined together and are transformed at the areas where the different sub-basins cross paths, as can be seen in the diagram below. For such a semi-distributed model to be successful, the hydrologist must divide the main drainage basin into smaller, homogenous sub-basins.

It is possible to use distributed models in which a specially adapted grid can be used to solve issues associated with the spatial variability of the different parameters that are used in the model. For distributed models, however, one of the main difficulties is the automatic construction of the drainage model. This main problem is caused due to the way in which each grid is placed one on top of the other. The drainage model is the key force behind the agglomeration of primary water levels that form part of the discharge that occurs at a drainage basin's outlet. Such a drainage model relies on the direction of slopes and gradients, which are calculated by DEMs. For several years now the analysis of relief using information technology (IT) methods has been the subject of the research projects of many hydrologists. In the 1970s, DEMs (they had another name during this period) were very rarely used because it was necessary to represent altitudes on a map and at this time this was not possible for DEMs [LAB 76]. Some time later [DEP 91] tools used to digitize contour lines were introduced, and these tools were seen as a leading development of that period in time. Progress made to date in the field of remote sensing has meant that DEMs have become increasingly available. However, several anomalies can also occur, such as the preferred direction in which the drainage flows, and the creation of endroheic basins, etc. It is, therefore, necessary to make more people aware of such drainage models and to get them to use them, as well as using the stream system that is also available. The different commands of GIS make it possible to use such models and include flow direction, flow accumulation and flow length, which are part of the library known as Hydro Arc View.

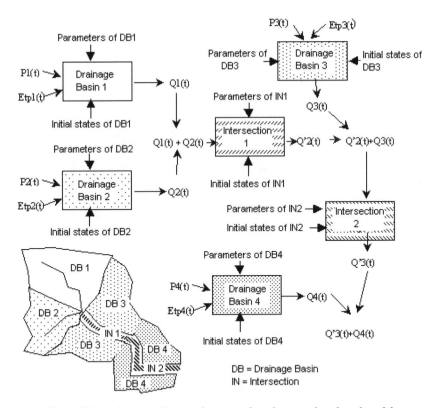

Figure 7.3. *A diagram showing the principles of a semi-distributed model*

From this introduction it is possible to see that in order to respond to the problems of today's society, hydrologists require precise information about water levels for a particular point on a drainage basin's outlet. These data are very rarely available, and in order to solve such issues, the hydrologist is required to use climatological data (mainly rainfall) and to recreate the relationship between rainfall and water levels that occurs within the drainage basin. This is, of course, subject to the geographical infomation that is available.

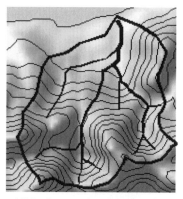

a) if the topography of each node in a grid network is known

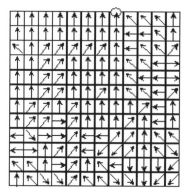

b) it is possible to calculate the direction in which each grid flows

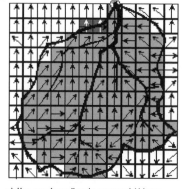

c) the contour line is correct if two grids in the network are adjusted

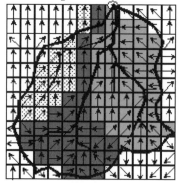

d) three drainage basins are detected instead of one

Figure 7.4. *Distributed models and the problems associated with the drainage model*

7.2. Pluviometry: a spatially continuous piece of geographical information

7.2.1. *Mathematically modeling precipitation*

Rainfall is a random phenomenon that possesses a certain amount of spatial continuity. Due to the data produced by rain gauges and pluviographs, it is possible to measure rainfall. Measurements recorded by remote sensing produce an image of the precipitation field. However, the accuracy of any indirect estimation is poor (see chapter on remote sensing) and such estimations can not be used to model the effect that rainfall has on water levels. At the moment, weather radars provide a relatively good estimation of the spatial distribution of rainfall. However, this type of measurement is not very widely available and is mainly used in developed countries. Furthermore, the measurements recorded only provide information about a short period of time. On a long-term basis, operational hydrology will be used to deal with specific data relating to rainfall. There are two different types of problems that can be associated with networks of rain gauges:

– estimating rainfall in a particular point for which there is no actual measurement available;

– evaluating average rainfall on a basin.

In order to solve the problems associated with the estimation methods, it is necessary to formalize the way in which rainfall is represented mathematically. Geostatistics were developed in France by G. Matheron [MAT 71] and his team at the École des Mines in Paris [DEL 78; JOU 80]. The École des Mines is a prestigious French engineering school in Paris. Some time later this idea of geostatistic modeling was successfully used in the analysis of precipitation fields [MAL 74; LAB 82; LEM 86]. Unlike geological data where only one observation is available, rainfall is a phenomenon that can be observed many times. According to the geostatistic vocabulary, rainfall is said to be a "climatologic data". Before going into further detail on the practical uses of the different estimation methods that exist, it is necessary to consider the theory behind them. This theory stems from research carried out by Obled [OBL 87], and from the works published by his team at the Institute of Mechanics in Grenoble, France [CRE 79] [TOU 81; LEB 84]. The research carried out by Obled may, at first, seem to be quite complex and may also seem to have little to do with the world of physics. This research has played an essential role in helping people effectively use the computing tools widely available today.

7.2.1.1. *Rainfall: a random function of order 2*

Where:

– R is a numerical value at a point $\vec{x}$ in a given domain D;

– ω is an event taken from a set Ω;

– P is a measurement of probability so that: $P(\Omega) = \int_{\Omega} P(d\omega) = 1$.

$R(\vec{x}, \omega)$ is a random function. For example $R(\vec{x}, \omega)$ is the total amount of rainfall that has been recorded by a rain gauge located at a point $\vec{x}$ in a particular region being investigated, noted as D. The period of investigation is noted as t and begins with an instant ω recorded during the nighttime Ω.

If an event ω_k is fixed, then the function $f_k(\vec{x}) = R(\vec{x}, \omega_k)$ is a trajectory of $R(\vec{x}, \omega)$. As far as rainfall is concerned, the function $f_k(\vec{x})$ can be represented by isohyetal curves measured between the instants of ω and $\omega + t$.

If we consider a particular fixed point, such as $\vec{x}_i$, the function $R_i(\omega) = R(\vec{x}_i, \omega)$ is said to be a random variable. In our study $R_i(\omega)$ is used to represent the probability distribution law of isolated rainfall that occurs at $\vec{x}_i$ over a period of time t.

$R(\vec{x}, \omega)$ is the random function average that occurs at the point $\vec{x}i$, the function $m(\vec{x}i)$ is expressed as follows:

$$m(\vec{x}i) = E\{R(\vec{x}i, \omega)\} = \int_{\Omega} R(\vec{x}i, \omega)P(d\omega)$$

Regarding rainfall, $m(\vec{x}i)$ represents the average of the total amount of rainfall that has fallen over a period of time, at a particular rain gauge located at $\vec{x}i$. The covariance function $C(\vec{x}i, \vec{x}j)$ (whenever it exists) can be expressed as follows:

$$C(\vec{x}i, \vec{x}j) = E\{R_i(\omega)R_i(\omega)\} - m(\vec{x}i)\,m(\vec{x}j) \qquad \text{with } C(\vec{x}i, \vec{x}j) = C(\vec{x}j, \vec{x}i)$$

A random function is said to be of order two if both the average and the covariance exist together. If a random function is of order two then a variance for each point $\vec{X}$ exists:

$$V(\vec{x}i) = \sigma^2(\vec{x}i) = C(\vec{x}i, \vec{x}i)$$

The correlation function $\rho(\vec{x}i, \vec{x}j)$ can be defined from the co-variance and variance functions:

$$\rho(\vec{x}i, \vec{x}j) = \frac{C(\vec{x}i, \vec{x}j)}{\sigma(\vec{x}i)\sigma(\vec{x}j)} \qquad \text{with } \rho(\vec{x}i, \vec{x}j) = \rho(\vec{x}j, \vec{x}i))$$

This formulation is quite similar to the representation that we have of precipitation. However, in order to continue with our research some more assumptions need to be made. Such assumptions include the fact that if it is possible to record annual rainfall each year (climatologic data), but the estimation of average rainfall has to be unique.

7.2.1.2. Homogenity and isotropy

If the correlation function is independent of the points $\vec{x}i$ and $\vec{x}j$ and only depends on the vector $\vec{h} = \vec{x}i - \vec{x}j$, the area being studied is said to be homogenous, and an anisotropic correlogram is present $\rho(\vec{h})$.

If the correlation function is independent of the points $\vec{x}i$ and $\vec{x}j$ and only depends on the normal function $h = \|\vec{h}\|$ of the vector $\vec{x}i - \vec{x}j$, the area being studied is said to be isotropic and an isotropic correlogram is present $\rho(h)$.

7.2.1.3. *Stationarity of order 2*

A random function is said to be stationary of order 2 if a correllogram exists and if the averages and variances of the function are present for each point $\vec{x}$:

$m(\vec{x}) = $ m (for each point $\vec{x}$ which is part of D),

$\sigma(\vec{x}) = \sigma$ (for all points $\vec{x}$ which are part of D),

$\rho(h) = \dfrac{C(\vec{x}i, \vec{x}i)}{\sigma^2}$ (for all points of $\vec{x}i$ and $\vec{x}i$ which are part of D).

The assumption that was made at the end of the previous section rarely occurs in practice, especially in the case of large drainage basins. A certain number of geographical factors, such as relief, distance from the sea, latitude, and exposure, highlight the fact that average rainfall is different in the area being investigated, i.e. area D. Using geographical information reduces all of these different factors into one pluviometric characteristic for which the hypothesis of stationarity is valid.

7.2.2. *Use with interpolation methods*

7.2.2.1. *General principles*

The problem with mathematical modeling comes to the fore when we want to recreate at any given point $\vec{x}o$, during an event ω , and for which the value $\overset{*}{R}(\vec{x}o, \omega)$ has not been recorded. The decision is then made to recreate $\overset{*}{R}(\vec{x}o, \omega)$ as a linear combination of observations that are recorded for the same period of rainfall on the network of rain gauges. This network is made up of a different number of points known as $\vec{x}i$:

$$\overset{*}{R}(\vec{x}o, \omega) = ao + \sum_{i=1}^{n} \lambda i R(\vec{x}i, \omega) \qquad [7.1]$$

The objective is to have an accurate estimation of the averages:

$$E\left\{ R(\vec{x}o, \omega) - \overset{*}{R}(\vec{x}o, \omega) \right\} = 0 \Rightarrow E\left\{ R(\vec{x}o, \omega) - ao - \sum_{i=1}^{n} \lambda i R(\vec{x}i, \omega) \right\} = 0$$

$$E\left\{ R(\vec{x}o, \omega) - \sum_{i=1}^{n} \lambda i R(\vec{x}i, \omega) \right\} = ao$$

$$ao = m(\vec{x}o) - \sum_{i=1}^{n} \lambda i m(\vec{x}i) \qquad [7.2]$$

The final objective is to recreate a given point by providing the best estimation of the averages by using the principle of the least squares:

$$\delta^2 = E\left\{ (R(\bar{x}o, \omega) - \overset{*}{R}(\bar{x}o, \omega))^2 \right\} = \min \qquad [7.3]$$

7.2.2.2. Autocorrelation

No hypothesis will be made about stationarity, and it is assumed that the unbiased equation [7.2] has already been examined and verified. The interpolation error can be represented by several different equations:

$$R(\bar{x}o, \omega) - \overset{*}{R}(\bar{x}o, \omega) = R(\bar{x}o, \omega) - ao - \sum_{i=1}^{n} \lambda_i R(\bar{x}i, \omega) \text{ but } m(\bar{x}o) = ao + \sum_{i=1}^{n} \lambda_i m(\bar{x}i)$$

$$R(\bar{x}o, \omega) - \overset{*}{R}(\bar{x}o, \omega) = R(\bar{x}o, \omega) - m(\bar{x}o) - \sum_{i=1}^{n} \lambda_i(R(\bar{x}i, \omega) - m(\bar{x}i))$$

$r(\bar{x})$ will be used to refer to the difference that exists between total rainfall and average rainfall ($r(\bar{x}) = R(\bar{x}, \omega) - m(\bar{x})$):

$$R(\bar{x}o, \omega) - \overset{*}{R}(\bar{x}o, \omega) = r(\bar{x}o, \omega) - \sum_{i=1}^{n} \lambda_i r(\bar{x}i, \omega) \qquad [7.4]$$

Equation [7.3] can, therefore, be represented as:

$$\delta^2 = E\left\{ (r(\bar{x}o, \omega) - \sum_{i=1}^{n} \lambda_i r(\bar{x}i, \omega))2 \right\} = \min$$

It is then possible to determine the n parameters of λ_i by canceling the n parameters that have been partially derived from δ^2 in relation to λ_i :

$$\delta^2 = E\{(r(\bar{x}o, \omega)2\} - 2\sum_{i=1}^{n} \lambda_i E\{r(\bar{x}o, \omega)r(\bar{x}i, \omega)\} + \sum_{j=1}^{n}\sum_{i=1}^{n} \lambda_i \lambda_j E\{r(\bar{x}j, \omega)r(\bar{x}i, \omega)\}$$

$$E\{r(\bar{x}o, \omega)r(\bar{x}i, \omega)\} = E\{ [R(\bar{x}o, \omega) - m(\bar{x}o)][R(\bar{x}i, \omega) - m(\bar{x}i)] \}$$

$$E\{r(\bar{x}o, \omega)r(\bar{x}i, \omega)\} = E\{R(\bar{x}o, \omega)R(\bar{x}i, \omega)\} - m(\bar{x}o)E\{R(\bar{x}i, \omega)\}..$$
$$... - m(\bar{x}i)E\{R(\bar{x}0, \omega)\} + E\{m(\bar{x}i)m(\bar{x}o)\}$$

$$E\{r(\bar{x}o, \omega)r(\bar{x}i, \omega)\} = E\{R(\bar{x}0, \omega)R(\bar{x}i, \omega)\} - m(\bar{x}i)m(\bar{x}o)$$

$$E\{r(\bar{x}o, \omega)r(\bar{x}i, \omega)\} = C(\bar{x}o, \bar{x}i)$$

where the equation:

$$\delta^2 = \sigma^2(\bar{x}o) - 2\sum_{i=1}^{n} \lambda_i C(\bar{x}o, \bar{x}i) + \sum_{j=1}^{n}\sum_{i=1}^{n} \lambda_i \lambda_j C(\bar{x}j, \bar{x}i)$$

and the derivatives can be written as follows:

$$\frac{\partial(\delta^2)}{\partial(\lambda_i)} = -2C(\bar{x}o, \bar{x}i) + 2\sum_{j=1}^{n} \lambda_j C(\bar{x}j, \bar{x}i)$$

The derivatives must have a value of zero. This means that for n values of λ_i, the following equation is valid:

$$\sum_{j=1}^{n} \lambda_j C(\vec{x}_j, \vec{x}_i) = C(\vec{x}_o, \vec{x}_i)$$

This equation can also be represented as a matrix:

$$\begin{vmatrix} C(\vec{x}_1, \vec{x}_1)C(\vec{x}_1, \vec{x}_2)...C(\vec{x}_1, \vec{x}_j)...C(\vec{x}_1, \vec{x}_n) \\ C(\vec{x}_2, \vec{x}_1)C(\vec{x}_2, \vec{x}_2)..C(\vec{x}_2, \vec{x}_j)..C(\vec{x}_2, \vec{x}_n) \\ ... \quad ... \quad ... \quad ... \quad ... \\ C(\vec{x}_i, \vec{x}_1)C(\vec{x}_i, \vec{x}_2)...C(\vec{x}_i, \vec{x}_j)...C(\vec{x}_i, \vec{x}_n) \\ C(\vec{x}_n, \vec{x}_1)C(\vec{x}_n, \vec{x}_2)..C(\vec{x}_n, \vec{x}_j)..C(\vec{x}_n, \vec{x}_n) \end{vmatrix} \begin{vmatrix} \lambda_1 \\ \lambda_2 \\ \vdots \\ \lambda_i \\ \lambda_n \end{vmatrix} = \begin{vmatrix} C(\vec{x}_1, \vec{x}_o) \\ C(\vec{x}_2, \vec{x}_o) \\ C(\vec{x}_i, \vec{x}_o) \\ C(\vec{x}_n, \vec{x}_o) \end{vmatrix} \Rightarrow [C_{i,j}][\lambda_i] = [C_{i,o}]$$

It is, therefore, possible to find all the values of λ_i by solving the following equation:

$$[\lambda_i] = [C_{i,j}]^{-1}[C_{i,o}]$$

It is also feasible to evaluate the variance of estimation δ^2 by using the equation:

$$\delta^2 = C(\vec{x}_o, \vec{x}_o) - \sum_{i=1}^{n} \lambda_i C(\vec{x}_i, \vec{x}_o)$$

Autocorrelation is not an operational method of interpolation; in order to successfully use the process of autocorrelation, the values of $C_{i,j}$ and $m(\vec{x}_i)$ need to be known. The values of $m(\vec{x}_o)$ and $C(\vec{X}_i, \vec{X}_0)$ also need to be known. However, this is never the case in practice. The actual interpolation methods used provide information on the values of $m(\vec{x}_o)$ and $C(\vec{x}_i, \vec{x}_o)$. This information can be obtained by following rather simplified hypotheses.

7.2.2.3. Covariance kriging

First of all it is assumed that that the random function is a stationary function of order 2:

$$m(\vec{x}) = m$$

$$\sigma^2(\vec{x}) = \sigma^2$$

$$C(\vec{x}_i, \vec{x}_j) = C(\vec{x}, \vec{x} + \vec{h}) = C(\vec{h})$$

the aim is to express $R(\vec{x}, \omega)$ using the following linear function:

$$\overset{*}{R}(\vec{x}_o, \omega) = \sum_{i=1}^{n} \lambda_i R(\vec{x}_i, \omega)$$

This equation differs from the previous ones because it is assumed that the value of a_o is zero. The equation is said to be unbiased:

$$a_0 = m(\vec{x}_0) - \sum_{i=1}^{n} \lambda_i m(\vec{x}_i)$$

and can be easily rewritten (since $m(\vec{x}) = m$ and $a_0 = 0$) as:

$$\sum_{i=1}^{n} \lambda_i = 1$$

The minimization of the variance of estimation leads to what is commonly known as a kriging system:

$$\sum_{j=1}^{n} \lambda_j C(\vec{x}_j, \vec{x}_i) - \mu = C(\vec{x}_0, \vec{x}_i)$$

This equation can also be written as a matrix:

$$
\begin{vmatrix}
C(\vec{x}_1, \vec{x}_1) C(\vec{x}_1, \vec{x}_2)...C(\vec{x}_1, \vec{x}_j)...C(\vec{x}_1, \vec{x}_n)\,1 \\
C(\vec{x}_2, \vec{x}_1) C(\vec{x}_2, \vec{x}_2)..C(\vec{x}_2, \vec{x}_j)..C(\vec{x}_2, \vec{x}_n)\,1 \\
C(\vec{x}_i, \vec{x}_1)\,C(\vec{x}_i, \vec{x}_2)...C(\vec{x}_i, \vec{x}_j)...C(\vec{x}_i, \vec{x}_n)\,1 \\
C(\vec{x}_n, \vec{x}_1) C(\vec{x}_n, \vec{x}_2)..C(\vec{x}_n, \vec{x}_j)..C(\vec{x}_n, \vec{x}_n)\,1 \\
1 \qquad 1 \quad ... \quad 1 \quad ... \quad 1 \quad 0
\end{vmatrix}
\begin{vmatrix}
\lambda_1 \\ \lambda_2 \\ \lambda_i \\ \lambda_n \\ \mu
\end{vmatrix}
=
\begin{vmatrix}
C(\vec{x}_1, \vec{x}_0) \\ C(\vec{x}_2, \vec{x}_0) \\ C(\vec{x}_i, \vec{x}_0) \\ C(\vec{x}_n, \vec{x}_0) \\ 1
\end{vmatrix}
$$

It is, therefore, necessary to develop a second hypothesis: the values of m, σ^2 and $C(\vec{h})$, which are normally defined from different events referred to as ω_k (climatological estimations), can now be defined from one spatial distribution of the estimations that are given for the entire domain of D. This hypothesis is known as the Ergodic hypothesis.

It is, therefore, assumed that:

$$m = E_\Omega \{ R(\vec{x}_i, \omega) \} = \int_\Omega R(\vec{x}_i, \omega) P(d\omega) = E_D \{ R(\vec{x}_i, \omega_k) \}$$

$$E_\Omega \{ R(\vec{x}_i, \omega) R(\vec{x}_j, \omega) \} - m^2 = E_D \{ R(\vec{x}_i, \omega_k) R(\vec{x}_j, \omega_k) \} - m^2 = \sigma^2 (1 - \rho(\vec{h})) = C(\vec{h})$$

The field average is evaluated by the following equation:

$$m \approx \frac{1}{n} \sum_{l=1}^{n} R(\vec{x}_i, \omega_k)$$

The covariance function $C(\vec{h})$ is evaluated in terms of distance Δh (in the case of an anisotropic function phases of azimuth are used instead). The couplets of $(\vec{x}_i, \vec{x}_j)$ are represented as follows:

$$\| \vec{x}_i - \vec{x}_j \| = h \pm \frac{\Delta h}{2}$$

where the equation is a number of t and from which it is possible to obtain one of the points of the covariance function:

$$C(h') \approx \frac{1}{t}\sum_{l=1}^{t} (R(\vec{x}_i,\omega_k)R(\vec{x}_j,\omega_k)) - m^2 \text{ with } h' = \frac{1}{t}\sum_{l=1}^{t}\left\| \vec{x}_i - \vec{x}_j \right\|$$

In order to obtain one of the points of the covariance function all that needs to be done is to adapt the values of $C(h')$ to a theoretical model $\overset{*}{C}(h')$ and to fill the $[C_{i,j}]$ and $[C_{0,i}]$ matrices with the corresponding $\overset{*}{C}(h)$ values.

What has just been introduced is the process of simple kriging, which can be used to recreate $R(\vec{x}_0,\omega) - \overset{*}{R}(\vec{x}_0,\omega)$, which has a zero average and for which the variance of δ^2 (the difference) is minimal. This minimal value can be represented by the following equation:

$$\delta^2 = C(\vec{x}0, \vec{x}0) + \mu - \sum_{i=1}^{n} \lambda_i C(\vec{x}i, \vec{x}o)$$

Such a case is very rarely dealt with in practice, and as a result the spatial estimation of m can sometimes be biased. Conversely, it is not strictly true that the variance of σ^2 is finite.

7.2.2.4. *Kriging under intrinsic hypotheses*

It is often necessary for an additional hypothesis to be created. This hypothesis is known as an intrinsic hypothesis, and in this hypothesis assumptions are made in relation to the increase in the values of $R(\vec{x}, \omega k) - R(\vec{x} + \vec{h}, \omega k)$ (which are stationary of order 2) for a given distance $\vec{h}$.

$$E_D\left\{ R(\vec{x}, \omega k) - R(\vec{x} + \vec{h}, \omega k) \right\} = m(\vec{x}) - m(\vec{x} + \vec{h}) = C_{te} = 0$$

The increased values have a zero average, and the value of $R(\vec{x}, \omega_k)$ tends to be stationary.

The constant variance for a distance $\vec{h}$ can be written as follows:

$$Var_D\left\{ R(\vec{x}, \omega k) - R(\vec{x} + \vec{h}, \omega k) \right\} = 2\gamma(\vec{h})$$

The intrinsic hypothesis can be summarized as follows: the values of the spatial increases are zero, and the variance of these increased values depends only on the vector $\vec{h}$ or on its module h.

$2\gamma(\vec{h})$ is referred to as a variogram. Regarding the intrinsic hypothesis, the variogram is associated with covariances and can be represented as follows:

$$2\gamma(\vec{h}) = Var_D\left\{ R(\vec{x}, \omega k) - R(\vec{x} + \vec{h}, \omega k) \right\}$$

$$2\gamma(\vec{h}) = E_D\left\{ \left[R(\vec{x}, \omega k) - R(\vec{x} + \vec{h}, \omega k) \right]^2 \right\} - \left[E_D\{R(\vec{x}, \omega k) - R(\vec{x} + \vec{h}, \omega k)\} \right]^2$$

$$2\gamma(\vec{h}) = E_D\left\{ \left[R(\vec{x}, \omega k) - R(\vec{x} + \vec{h}, \omega k) \right]^2 \right\} - 0$$

$$2\gamma(\vec{h}) = E_D\left\{ [R(\vec{x}, \omega k)]^2 \right\} - 2E_D\left\{ R(\vec{x} + \vec{h}, \omega k)R(\vec{x}, \omega k) \right\} + E_D\left\{ \left[R(\vec{x} + \vec{h}, \omega k) \right]^2 \right\}$$

However:

$$E_D\left\{ [R(\vec{x}, \omega k)]^2 \right\} - m^2 = E_D\left\{ \left[R(\vec{x} + \vec{h}, \omega k) \right]^2 \right\} - m^2 = C(0)$$

$$2E_D\left\{ R(\vec{x} + \vec{h}, \omega k)R(\vec{x}, \omega k) \right\} - 2m^2 = 2C(\vec{x}, \vec{x} + \vec{h}) = 2C(\vec{h})$$

Therefore:

$$2\gamma(\vec{h}) = ED\left\{ [R(\vec{x}, \omega k)]^2 \right\} - m^2 - 2ED\left\{ R(\vec{x} + \vec{h}, \omega k)R(\vec{x}, \omega k) \right\}..$$
$$... + 2m^2 + ED\left\{ \left[R(\vec{x} + \vec{h}, \omega k) \right]^2 \right\} - m^2$$

$$\gamma(\vec{h}) = C(0) - C(\vec{h})$$

The initial system can therefore be transformed into:

$$\overset{*}{R}(\vec{x}_o, \omega) = \sum_{i=1}^{n} \lambda_i R(\vec{x}_i, \omega) \text{ with } \sum_{j=1}^{n} \lambda_j \gamma(\vec{x}_j, \vec{x}_i) - \mu = \gamma(\vec{x}_0, \vec{x}) \text{ et } \sum_{i=1}^{n} \lambda_i = 1$$

and can also be represented as a matrix:

$$
\begin{vmatrix}
\gamma(\vec{x}1, \vec{x}1)\,\gamma(\vec{x}1, \vec{x}2)...\gamma(\vec{x}1, \vec{x}j)...\gamma(\vec{x}1, \vec{x}n)\,1 \\
\gamma(\vec{x}2, \vec{x}1)\,\gamma(\vec{x}2, \vec{x}2)...\gamma(\vec{x}2, \vec{x}j)...\gamma(\vec{x}2, \vec{x}n)\,1 \\
\gamma(\vec{x}i, \vec{x}1)\,\gamma(\vec{x}i, \vec{x}2)...\gamma(\vec{x}i, \vec{x}j)...\gamma(\vec{x}i, \vec{x}n)\,i \\
\gamma(\vec{x}n, \vec{x}1)\,\gamma(\vec{x}n, \vec{x}2)..\gamma(\vec{x}n, \vec{x}j)..\gamma(\vec{x}n, \vec{x}n)\,i \\
1 \quad\quad 1 \quad ... \quad 1 \quad ... \quad 1 \quad 0
\end{vmatrix}
\begin{vmatrix}
\lambda 1 \\ \lambda 2 \\ \lambda_i \\ \lambda n \\ \mu
\end{vmatrix}
=
\begin{vmatrix}
\gamma(\vec{x}1, \vec{x}0) \\ \gamma(\vec{x}2, \vec{x}0) \\ \gamma(\vec{x}i, \vec{x}0) \\ \gamma(\vec{x}n, \vec{x}0) \\ 1
\end{vmatrix}
$$

So that the variogram can be used in practice, an experimental variogram needs to be developed:

$$2\gamma(\vec{h}) = ED\left\{ \left[R(\vec{x}, \omega k) - R(\vec{x} + \vec{h}, \omega k) \right]^2 \right\}$$

This equation is created by evaluating the distance Δh (in the case of an anisotropic function phases of azimuth are used instead). The couplets of $(\vec{x}i, \vec{x}j)$ are represented as follows:

$$\left\| \vec{x}i - \vec{x}j \right\| = h \pm \frac{\Delta h}{2}$$

where the equation is a number of t and from which it is possible to obtain the variogram:

$$2\gamma(h') \approx \frac{1}{t}\sum_{l=1}^{t} (R(\vec{x}_i,\omega_k)R(\vec{x}_j,\omega_k)) \text{ with } h'=\frac{1}{t}\sum_{l=1}^{t}\left\|\vec{x}_i-\vec{x}_j\right\|$$

$\gamma(h')$ is adapted to the theoretical model $\overset{*}{\gamma}(h')$ and the matrices $[\Gamma_{i,j}]$ and $[\Gamma_{0,i}]$ are completed with the corresponding values of $\overset{*}{\gamma}(h)$.

7.2.3. Examples of mapping precipitation

After this short introduction to the theory that is associated with the spatial distribution of precipitation the next part of this chapter will focus more on actual case studies. These case studies are taken from recent research that has been carried out by the University Sophia Antipolis in Nice, France. The university undertook this research on behalf of the Algerian National Agency for Hydraulic Resources [ASS 04]. The aim of this research was to create maps showing the monthly rainfall totals for the period September 1965 to August 1995, in other words 360 maps would be created. These maps would provide useful information that could be used as input data in a model used to analyze rainfall and water levels.

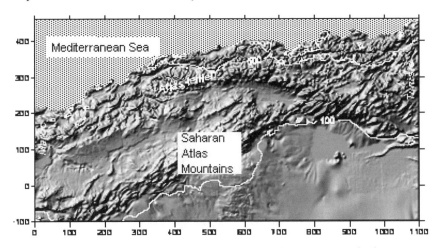

Figure 7.5. *The general characteristics of the study area, i.e. North Algeria*

Northern Algeria is 1,100 km long (from the Algerian-Moroccan border through to the Algerian-Tunisian border) and is 550 km wide (from the Mediterranean to the Saharan Atlas Mountains). The mountains are very steep and face WSW-ENE, meaning that there is a high level of precipitation anisotropy. Average annual rainfall ranges from less than 100 mm to the south of the Atlas Mountains to more than 500 mm in a large part of coastal areas, and exceeds 1,000 mm in Kabylie [1].

This contrast in rainfall levels can also be applied to the monthly averages. The statistical distribution of monthly rainfall is not uniform and cannot be represented by Gaussian theory. Rainfall for a particular month and for a particular year can only be modeled by a random non-Gaussian function, which is both anisotropic and non-stationary. The example of Algeria focuses on the difficulties that are associated with the automatic mapping of the precipitation fields.

7.2.3.1. Identification of the random variable and the normalization of data

Research has shown that irrespective of the season and of the location, the statistical distributions of monthly rainfall produce a positive skew. However, for the rainy months (see Figure 7.6) the distributions were seen as being normal square roots.

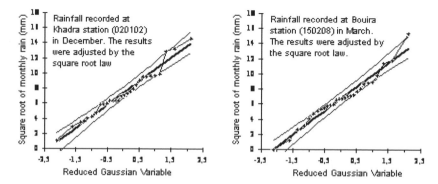

Figure 7.6. *Examples of total monthly rainfall distribution for the rainy months*

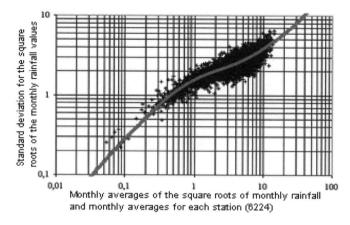

Figure 7.7. *The relationship between the square roots of monthly rainfall averages and the standard deviation associated with the square roots of these averages*

The distribution of monthly rainfall is, therefore, expressed as square roots and is made up of the following parameters:

– $\overline{\sqrt{P}}$ is the average of the square roots of the monthly rainfall (the square of $\overline{\sqrt{P}}$ gives the monthly rainfall median, Pmed);

– $\sigma_{\sqrt{P}}$ is the standard deviation of the square roots of the monthly rainfall.

The adjustments were applied to every rain gauge that formed part of the rain gauge network. This network held data that provided information on observations that had been measured for a period of at least 20 years during the period from September 1965 to August 1995. Depending on the month being studied, the number of stations used to take rainfall measurements varied from 499 to 535. Distribution of rainfall and average rainfall are dependent on one another. These two parameters are closely linked so that any differences that exist between the two can be explained by uncertainties that are associated with the estimations of the values of $\sigma_{\sqrt{P}}$. Figure 7.7 shows that it is possible to obtain $\sigma_{\sqrt{P}}$ from the following equation:

$$\sigma_{\sqrt{P}} \approx 0.2188\overline{\sqrt{P}} + 1.5852(1 - e^{-\overline{\sqrt{P}}/0.5489})$$

The variable U(P) can, therefore, be considered as a random reduced centered Gaussian variable:

$$U(P) = \frac{\sqrt{P} - \overline{\sqrt{P}}}{0.2188\overline{\sqrt{P}} + 1.5852(1 - e^{-\overline{\sqrt{P}}/0.5489})}$$

7.2.3.2. *Mapping the averages*

For any given month the square root of rainfall is a random Gaussian variable with an average of $\overline{\sqrt{P}}$. Only one value is known for each rain gauge and some doubt is associated with this value. The doubt arises because the values of rain gauges vary depending on how long they have been in active use to measure rainfall. This parameter is non-stationary, increases with altitude and decreases with distance from the sea. As the square of $\overline{\sqrt{P}}$ represents the median values of rainfall, these values will be the first to be represented on a map.

The median values of monthly rainfall are associated with the following factors: the altitude at which the rain gauge is located (referred to as Z, measured in meters); the longitude, which is measured by the Lambert x in kilometers; and distance from the sea (referred to as d and measured in kilometers). Distance from the sea has an influential role on the median values of monthly rainfall for the first few kilometers and for this reason we have decided to use distance from the sea as an exponential decline, with the parameter k being a regional optimum for a particular month. The parameter k refers to the speed at which the effect of distance from the sea decreases. The regression used to represent the median values is as follows:

$$Pmed = a*e^{-d/k} + b*ze^{-d/k} + c*Zliss + d*xe^{-d/k} + e*xze^{-d/k} + C^{te} + \varepsilon$$

Table 7.1 provides information on the monthly values of parameter k (the speed at which the effect of distance from the sea decreases) and of other regression parameters. M refers to the month of the year in which the values were measured.

M	K	$e^{-d/k}$	$ze^{-d/k}$	Z	$xe^{-d/k}$	$xze^{-d/k}$	Cte	R
1	43	−2.1E+01	1.1E−0.1	−4.6E−03	1.5E−01	−1.8E−05	16	0.89
2	47	−6.4E+00	1.5E−01	−4.0E−03	1.1E−0.1	−9.6E−05	14	0.86
3	43	−5.7E+00	1.3E−0.1	5.1E−04	1.0E−01	−2.4E−05	18	0.85
4	60	2.0E+00	6.1E−02	1.2E−03	6.6E−02	−5.8E−06	10	0.84
5	144	2.1E+01	5.9E−03	8.5E−03	1.7E−02	8.5E−06	−5	0.73
6	2959	8.4E+01	−4.3E−02	4.5E−02	7.7E−03	3.6E−06	−82	0.71
7	3337845	2.3E+04	−1.5E+01	1.5E+01	1.0E−03	1.8E−06	−23337	0.64
8	44005838	8.4E+05	−9.6E+02	9.6E+02	2.5E−03	6.6E−06	−842362	0.78
9	185	2.3E+01	−2.6E−02	2.3E−02	2.7E−02	1.2E−05	−18	0.79
10	33	−9.2E+00	2.9E−02	8.1E−04	9.8E−02	2.8E−05	15	0.88
11	39	2.3E+00	1.1E−01	3.2E−03	1.1E−01	−9.0E−05	12	0.87
12	39	−1.9E+01	1.4E−01	−4.5E−03	1.7E−01	−6.6E−06	16	0.90

Table 7.1. *Monthly values of k and regression coefficients*

From the table it can be seen that the median values for summer rainfall (from May to September) are not greatly influenced by topography, the multiple correlation coefficient of R is less than 0.8. For the months of May to August distance from the sea has a large effect on the median values of precipitation ($k>3,000$ km). The median rainfall value for the month of December can be calculated as follows:

$$Pmed_{12} \gg -19.20*e^{-d/43} + 0.1361*ze^{-d/43} - 0.005*Zliss...$$

$$... + 0.169*xe^{-d/43} - 0.000007*xze^{-d/43} + 15.81$$

The following equation:

$$Pmed = a*e^{-d/k} + b*ze^{-d/k} + c*Zliss + d*xe^{-d/k} + e*xze^{-d/k} + C^{te} + \varepsilon$$

is an approximate equation due to the existence of the local corrective term ε. The local corrective term ε possesses a zero average and the value of this term does not depend on distance from the sea, altitude, etc. Regarding the intrinsic hypothesis, we are actually dealing with the same conditions in which the process of kriging would be used. In order to plot this local corrective term on a map, more research needs to be carried out on its variogram.

As can be seen in Figure 7.8, variograms tend to be spherical. They are also associated with what is known as a nugget effect, and tend to be anisotropic in nature. Variograms have ranges that exist in a ratio of 1:2 for angles of both 5° and 95°.

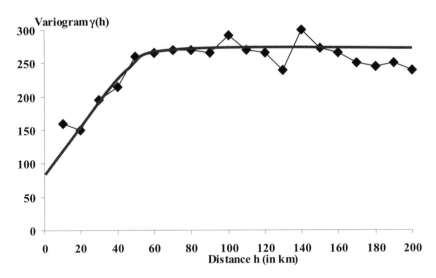

Figure 7.8. *Variogram of the residual rainfall for the month of December, flowing in a North-South direction at an angle of 95°*

Surprisingly, a nugget effect of 80 mm^2 exists for the month of December. This can be quite easily explained because the median values are only estimations taken from a reduced sample of rain gauges (of between 20 and 30 years).

For a common rain gauge whose true median rainfall is 45 mm ($\sqrt{P}$ =6.7 and $\sigma_{\sqrt{P}}$ =3.1) the median's variance of estimation increases to 75 mm^2 and this is due to the fact that only a sample of 25 years were used. A local variance, which is the same size as the nugget effect (in other words 80 mm^2) also exists for the entire zone that is being studied. The interpolation method of kriging is applied to the regression residuals on a monthly basis. This grid of residuals is then added to the initial map drawn up to show the estimations of rainfall levels. These estimations were made according to the geographical information that was available. As a result it is, therefore, possible to create the final map showing the median rainfall level for the month of December.

As can be seen in Figure 7.9, 12 different maps are created for a particular month and they show monthly median rainfall levels. The maps were created by taking geographical information into consideration, and by respecting the hypotheses that were developed to create the different kriging equations.

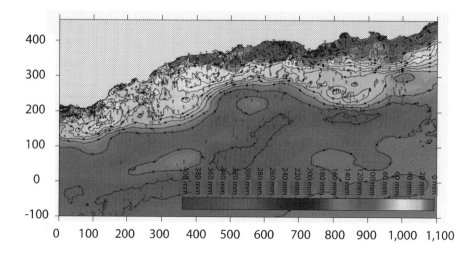

Figure 7.9. *Example of a map showing median rainfall values
for the month of December (see color section)*

This method of creating 12 maps showing median monthly rainfall levels has been used for a long time. For the purpose of our research we have focused on work that was carried out by geographers who were based in the French city of Strasbourg [SCH 77]. At that time maps had to be created manually. The technique that is used today was first used in France in 1982 [LAB 82]. This modern technique was quickly adopted, updated, and improved upon by the scientific community [LAB 84; JOR 86; BEN 87]. Similar approaches were then used to measure other climatic parameters, such as temperature, which is greatly affected by topography [CAR 94]. Up until the 1990s, rainfall levels were estimated according to the topographic factors that were used in the estimation process [LAB 95]. If too many topographic factors are used to estimate rainfall levels, there is also an increase in the number of redundant independent variables that are used in the estimation process. If these topographic factors are used to estimate rainfall levels in large areas (with a surface area of more than 500 km^2), there is less focus on site effect gauges. There is, however, more focus on the variations of the different topographic factors used in the measurement process. Another area for consideration involves the almost universal choice of the linear regression model. The altimetric gradient of rainfall is not constant all over the world and because of this it has become necessary to use non-linear equations. For some years now it has been clear that there is a limited number of factors that affect rainfall levels; these variables include altitude, distance from the sea, and distance from the crest of a mountain, etc [BER 00; KIE 01; DJE 01]. There is a general relationship that exists between rainfall levels and relief [LAB 02], and between rainfall levels and distance from the sea [ZAH 00].

The interpolation of the different statistical parameters characteristic of rainfall, which are used as geographical information, is a method that is recognized and used globally. The same is also true for the method used to calculate the different averages of the rainfall levels (such as arithmetic average and median), and for the method used to calculate dispersion parameters (or gradex), as well as the method used to work out the decennial total rainfall.

7.2.3.3. *Plotting rainfall levels on a map for a particular month*

For the long-term stations (those which have been used more than 20 times to provide information on rainfall levels for a given month, and in our case this is the month of December), it is possible to evaluate the median value and with this value it is possible to generate a map showing the rainfall levels for that particular month. If the median value is known, it is possible to estimate the value of $\sigma_{\sqrt{P}}$ by using the following equation:

$$\sigma_{\sqrt{P}} \approx 0.2188\overline{\sqrt{P}} + 1.5852(1 - e^{-\overline{\sqrt{P}}/0.5489})$$

X	y	Name	Code	Calculated Median	Median plotted on map	Rainfall Dec 94	Gaussian (U) variable
412.3	72.8	BRIDA	10101	14.67		0	-1.58
426.5	79.8	SEBGAG	10102		9.81	0.2	-1.19
459.7	116.3	CENTER	10204		8.11	0	-1.29
430.2	111.6	SIDI BOUZID	10205	8.56		0	-1.32
444.7	91.1	GUELTAT SIDI SAAD	10208		6.74	3.5	-0.34
464.5	177.8	AFLOU SECTOR	10502	3.10		0	-0.92
403.6	214.5	ZMALET EL AMIR AEK	10701	28.28		12.4	-0.65
434.5	234.5	AIN BAADJ	10703	18.31		11.4	-0.36
465.9	213.3	RECHAIGA	10704		9.35	3.9	-0.48
470.6	229.6	KSAR CHELLALA SIDI BOUDAOUD	10706	7.72		10.6	0.22

Table 7.2. *An extract taken from the observations that were recorded in December 1994*

In this section we will focus on rainfall levels for the month of December 1994, which can be seen in Table 7.2. The same type of table was generated for the other 359 months that were used in the study that took place between September 1965 and August 1995. For each measurement that was recorded in the month of December 1994 the median value for rainfall was known. This value was known, irrespective of whether the measurement was taken from a long-term station or from a station that was only used for a particular month.

For example, if we take a closer look at the long-term station of Rechaïga, we can see that the total level of rainfall for the month of December 1994 was 11.4 mm. The median rainfall value for the month of December (and which was measured over a longer period of time prior to the 1994 recording) was 18.31 mm. This means that the rainfall levels for December 1994 can also be seen as the creation of a random variable having a square-root normal distribution and is made up of the following parameters:

$$\overline{\sqrt{P}} = \sqrt{18.31} = 4.28$$

$$\sigma_{\sqrt{P}} \approx 0.2188\,\overline{\sqrt{P}} + 1.5852(1 - e^{-\overline{\sqrt{P}}/0.5489})$$

$$\sigma_{\sqrt{P}} \approx 0.2188 * 4.28 + 1.5852 * (1 - e^{-4.28/0.5489}) = 2.52$$

The rainfall levels for the month December 1994, therefore, correspond to a reduced Gaussian variable which is made up of:

$$u = \frac{\sqrt{11.4} - 4.28}{2.52} = -0.36$$

At the short-term station of Sebgag (which was used to record less than 20 observations for the month of December), we can see that the total level of rainfall for the month of December 1994 was 0.22 mm and that the estimated median for the same month was 9.81 mm, according to the map that was generated. This means that the rainfall levels for December 1994 can also be seen as the creation of a random variable which has a square-root normal distribution and is made up of the following parameters:

$$\overline{\sqrt{P}} = \sqrt{9.81} = 3.13$$

$$\sigma_{\sqrt{P}} \approx 0.2188\,\overline{\sqrt{P}} + 1.5852(1 - e^{-\overline{\sqrt{P}}/0.5489})$$

$$\sigma_{\sqrt{P}} \approx 0.2188 * 3.13 + 1.5852 * (1 - e^{-3.13/0.5489}) = 2.27$$

The rainfall levels for the month of December 1994, therefore, correspond to a reduced Gaussian variable made up of:

$$u = \frac{\sqrt{0.2} - 3.13}{2.27} = -1.19$$

These reduced Gaussian variables are characteristic of the relative intensity of the rainfall that fell in these areas during 1994. With these variables and values in mind, it is possible to carry out a process of spatial interpolation. When these Gaussian or U variables are created, they have a zero average and a variance level of one. Therefore, it is possible to study the structure and the function of each structure of the individual variables by creating a correlogram representing the total amount of rainfall that fell during the month of December 1994.

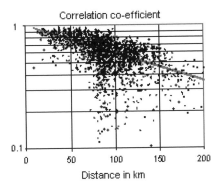

Figure 7.10. *Correlogram showing rainfall levels in the study zone used as part of research in Algeria*

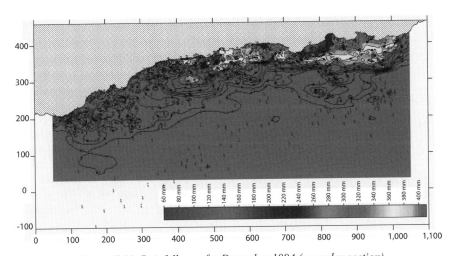

Figure 7.11. *Rainfall map for December 1994 (see color section)*

The different values present in the Figure 7.11 shown above have all been subjected to the process of simple kriging. There is no nugget effect because the exact values of the rainfall levels for December 1994 are known, whereas the median values are only estimations. These different values are all anisotropic and have two values when the water flowing on the surface is recorded at an angle of 5° rather than at 95°. These different values are also represented on the correlogram.

Given the information on the reduced Gaussian variables (U variables), it then becomes possible to create the precipitation map for the month of December 1994 as the value of $\overline{\sqrt{P}}$ is known and as a result:

$$P_{12-94} = \max\left[\ \sqrt{\overline{P}} + U*\left\{0.2188\sqrt{\overline{P}} + 1.5852(1 - e^{-\sqrt{\overline{P}}/0.5489})\right\};0\ \right]^2$$

(The maximum function [;] means that it is possible to work out rainfall levels.)

This method was also used to create the 360 monthly precipitation maps for the period between September 1965 and August 1995. Figure 7.11 is an example of one of these maps that was created.

7.2.3.4. *A summary of rainfall mapping*

It is not possible to highlight the exact effect that relief or distance from the sea has on rainfall levels for a particular month. However, these two factors influence the average monthly rainfall values. By applying the process of interpolation to the reduced variables it is possible to generate maps that are coherent with the maps that show information on monthly median rainfall levels.

The application of the method of interpolation to the reduced variables has improved estimations of average rainfall levels in the drainage basins. This improvement is important for drainage basins whose rain gauges are located at the bottom of a valley. If the relief of the local area is not taken into consideration, then rainfall levels tend to be underestimated and this underestimation means that the variables influencing rainfall levels need to be modified. Before carrying out this study, which focuses on monthly rainfall levels, the model known as LOIEAU was tested on 55 drainage basins from which average rainfall was recorded on rain gauges that were found either in the basin or near it. The LOIEAU was derived from the GR2M model. After slightly adjusting the values of the coefficients, the average of the 55 X1 coefficients was 1.13 and its standard deviation was 0.69. By taking measurements of rainfall levels that were recorded by rain gauges located only in the drainage basins, the average of the coefficients was very similar at 1.14, however, the standard deviation fell to a value of 0.47. A large part of the variance of the parameters used to adjust the values of the coefficients was used to compensate for errors that were created whenever rainfall levels were estimated [LAB 03]. Similar results were recorded in the French regions of Corsica and Provence by the CEMAGREF [LAV 97; ASS 00]. The CEMAGREF is a public and agricultural research institute that exists in France.

7.3. The problems associated with recording rainfall and average spatial rainfall

In the previous section we saw that it is possible to estimate rainfall for any part of the country by using approximate measurements, such as radar and remote sensing. However, using these methods is a relatively modern process and they are seldom used as they often lead to inaccurate results. The most common method used is interpolation, which is applied to isolated rainfall, as was described in the previous section. Hydrologists are more interested in the average spatial rainfall that falls in a particular drainage basin rather than levels of rainfall. In order to show

how it is possible to record average spatial rainfall in a drainage basin, two different observations can be made: first, the random function is stationary (which is the case for small urban drainage basins), and second, rainfall levels generally vary and this is the case for larger drainage basins.

7.3.1. *Reduction coefficient*

7.3.1.1. *Definition and traditional methods*

In order to evaluate the volume of water flowing on a surface, it is necessary to know the exact amount of rainfall that has fallen and has generated this flow of water. The level of isolated rainfall for a particular area tends to be known (by the use of a rain gauge), and not the average rainfall that has fallen on the surface of the drainage basin. The term K will be used to refer to the reduction coefficient. K is used to represent the relationship between the average rainfall that has fallen on a surface P_{BV}, and the isolated rainfall that has fallen with the same frequency F.

The value of the coefficient K depends on the surface of the drainage basin, on the frequency of rainfall, on the rain itself, as well as on the spatial distribution of the rainfall. K is only valid if the isolated rainfall with frequency F is the same at each point in the drainage basin. We have made the assumption that rain is a random stationary function. This means that the frequential distribution of the isolated rainfall is the same at each point in the drainage basin being investigated.

For many years now, studies that have been carried out on a regional basis to elucidate K coefficient have remained too empirical. One of the first solutions resulting from all this research was that developed by Brunet-Moret and Roche [BRU 66]. Their solution takes into consideration measurements recorded over both short and long periods of time.

This technique, and others developed after it, evaluate and examine the spatial average rainfall for the drainage basin for all recorded rainfall. These were long and tedious processes, and they could not be used to extrapolate results that provided information on the surface area of the drainage basin, duration of rainfall, and the frequency of rainfall.

7.3.1.2. *Geostatistic approaches*

The geostatistic approach, which was developed by Laborde [LAB 86] and then Lebel [LEB 88], relies on the simple geostatistic observations that were mentioned earlier in this section. The advantage of using this method is that it is no longer necessary to calculate average rainfall. If the statistical distribution of the rainfall and its structure can be worked out by using a variogram or a correlogram, it then becomes possible to evaluate the reduction coefficients for all frequencies of rainfall and for all drainage basins.

Rainfall that falls in the drainage basin needs to be considered as being the result of a random stationary function with an order of 2:

$$m(\bar{x}i) = m \quad V(\bar{x}i) = \sigma^2(\bar{x}i) = C(\bar{x}i, \bar{x}i) = \sigma^2 = V \quad C(\bar{x}i, \bar{x}j) = \sigma^2 - \gamma(\bar{x}i, \bar{x}j)$$

The spatial average of the rainfall in a drainage basin with area S is a random variable $Rs(\omega)$ so that:

$$Rs(\omega) = \frac{1}{S} \int_S R(\omega, \bar{x}) d\bar{x}$$

It is possible to show that the average ms of $Rs(\omega)$ is the same as the average of the rainfall m: $ms = E[Rs(\omega)] = m$. The variance σ_s^2 of average rainfall in a drainage basin can be expressed as follows:

$$\sigma_S^2 = \sigma 2 - \frac{1}{S^2} \int_S \int_S \gamma(\bar{x}i - \bar{x}j) \, d\bar{x}i \, d\bar{x}j$$

If isolated rainfall is distributed according to the same law that governs average rainfall across a drainage basin, then it is possible to calculate average rainfall and variance of average rainfall occurring in a drainage basin. Average rainfall and the variance of average rainfall can be calculated from the variogram of the isolated rainfall and from the domain S (the drainage basin which is used to calculate the reduction coefficient). Variograms can often be modeled by using an exponential equation:

$$\frac{\gamma(\bar{x}i - \bar{x}j)}{\sigma^2} = \frac{\gamma(\bar{h})}{\sigma^2} = 1 - e^{-h/p}$$

The variance σ_S^2 of average rainfall in a drainage basin can be expressed as:

$$\sigma_S^2 = \sigma^2 \, g(S, p)$$

where $g(S, p)$ is only a function of the area and shape of the drainage basin and the parameter p describes the spatial structure of rainfall. If the drainage basin is considered as being a square, the following equation is true:

$$g(S, p) = 1 - \frac{0.245\sqrt{S}}{p}$$

Measuring the reduction coefficient involves calculating the average and variance of isolated rainfall (depending on the station that has been in use for the longest period of time). It also involves determining which parameter p should be used to represent the spatial structure of rainfall (depending on the correlogram or variogram that has been created from different series of data of varying length; these series of data are recorded by the rain gauges). From this information, it is possible to work out the average, and variance of the distribution of average rainfall for a particular period of time. It is also possible to generate probability laws that can be used to work out average rainfall and reduction coefficients.

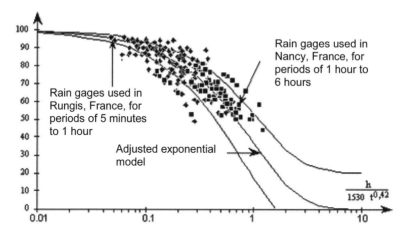

Figure 7.12. *The function of the structure of rainfall in the regions of Paris and Lorraine in France*

For example, we analyzed the correlation coefficients (referred to as r) that were produced by Bergaoui and Desborde in Rungis, near Paris [BER 87]. Bergaoui and Desborde used a set of rain gauges that recorded data over different time periods, ranging from 5 minutes up to 1 hour. Regarding the correlation coefficients for the Urban District of Nancy, the rain gauges that were used also recorded data over different time periods, from 1-6 hours. The correlograms produced can be adjusted so that they can be used with exponential models and so that the parameters of p vary as a function of duration of time. The following equation is, therefore, used to describe the entire set of observations:

$$r(h,t) = e^{-\frac{h}{1,530\ t^{0.42}}}$$

(h is expressed in meters and t is expressed in minutes)

For a drainage basin whose shape is considered as being square we have seen that:

$$\frac{\sigma_S}{\sigma} = 1 - \frac{0.245\sqrt{S}}{p},$$

which leads to:

$$\frac{\sigma_s}{\sigma} = 1 - \frac{0.245\sqrt{S}}{1,530\ t^{0.42}}$$

If average rainfall present in the drainage basin follows a Gumbel law, the following is true:

$$F(x_S) = e^{-e^{-(x-x_{oS})/g_S}}$$

with $g_S = 0.78\sigma_S$ and $x_{oS} = m - 0.577g_S$ or:

$$x_{S_F} = x_{oS} + g_S\left[- Ln\left\{- Ln(F)\right\}\right]$$

Whenever the value of the frequency is closer to one, the second variable is predominant and the equation can then be written as:

$$x_{S_F} \approx g_S\left[- Ln\left\{- Ln(F)\right\}\right]$$

$$x_{S_F} \approx 0.78\sigma_S\left[- Ln\left\{- Ln(F)\right\}\right]$$

This is also true for isolated rainfall:

$$x_{p_F} \approx 0.78\sigma\left[- Ln\left\{- Ln(F)\right\}\right]$$

and for more uncommon frequencies, the reduction coefficient K (S,t) becomes:

$$K\ (S,t) = x_{F_S}/x_{F_P} = \sigma_S / \sigma$$

$$K(S,f) = 1 - \frac{\sqrt{S}}{36t^{0.42}}$$

where S is measured in kilometers square and t is measured in hours.

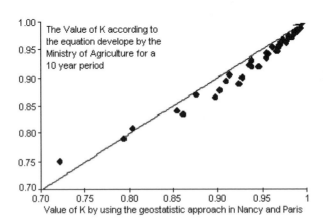

Figure 7.13. *A comparison of the results from traditional and geostatistic approaches*

This equation is very similar to the one that was generated by the Ministry of Agriculture [GAL 82], which was drawn up to measure rainfall levels that were recorded over a 10 year period in the region of Paris:

$$K(S,t) = 1 - \frac{\sqrt{S}}{30t^{0.33}}$$

This equation was developed empirically, after having examined K values by using traditional methods for all types of surfaces and for all durations of time. As can be seen in Figure 7.13, there is a strong correlation between the duration of observation from 1 hour to 20 hours and drainage basins with an area of 1 to 100 km^2.

7.3.1.3. Perspectives

The geostatistic modeling of rainfall has paved the way for research to be carried out on reduction coefficients. Factors that were once difficult to integrate into research, such as time, frequency, or shape of the drainage basin, can now be integrated into this new research. More research still needs to be carried out in this field, even though it is now a commonly used method to measure rainfall levels.

7.3.2. Average spatial rainfall

The aim here is to evaluate the rainfall that falls within a drainage basin as a whole. This is made possible by studying the spatial average for each period of rainfall without having to apply the process of interpolation to rainfall recorded at each different measuring site located in the drainage basin. This is a very old problem that different generations of geographers and hydrologists have had to resolve by using Thiessen polygons or isohyetal methods. In the next section we will see how the geostatistic approach for measuring rainfall is used to help geographers and hydrologists overcome this problem.

7.3.2.1. Rainfall can be derived

This derivation process can only occur under conditions where stationarity is not possible. We have already seen in this chapter that it is possible to interpolate rainfall $\overset{*}{R}(\vec{x}_o, \omega)$ at any given point in a drainage basin without having to rely on measurements of autocorrelation by using the equation:

$$\overset{*}{R}(\vec{x}_o, \omega) = a_o + \sum_{i=1}^{n} \lambda_i R(\vec{x}_i, \omega)$$

$$a_o = m(\vec{x}_o) - \sum_{i=1}^{n} \lambda_i m(\vec{x}_i)$$

$$\left[\lambda_i \right] = \left[C_{i,j} \right]^{-1} \left[C_{i,o} \right]$$

In these equations the parameters of a_0 and λ_i are independent of the observations that are made during a particular event (period of rainfall) ω. At each measurement site within the drainage basin, rainfall is treated as being a linear combination of the observations that have been made. This means that average spatial rainfall is also a linear combination of the observations that have been made in the drainage basin. In order to evaluate the accuracy of each measurement site as far as the estimation of spatial averages in the drainage basin are concerned, the values of the field averages $m(\vec{x})$, the field variances $\sigma^2(\vec{x})$ and the correlogram $\rho(h)$ must be known.

$$\text{(with } C(\vec{x}i, \vec{x}j) = \sigma(\vec{x}i) \; \sigma(\vec{x}j) \;\; \rho(\vec{x}i - \vec{x}j) \text{)}.$$

The parameters a_{0_k} and λ_{i_k} are then calculated for a total of n points $\vec{x}0_k$, which are part of a regular grid and which can be found inside the contours of a drainage basin. When the averages of these parameters are known it is then possible to calculate the amount of rainfall that falls in the drainage basin $R(\omega)_{BV}$ for an event ω.

$$\bar{a}o = \frac{1}{n} \sum_{k=1}^{n} a_{0_k}, \quad \bar{\lambda}i = \frac{1}{n} \sum_{k=1}^{n} \lambda_{i_k}$$

$$R(\omega)_{BV} = \bar{a}o + \sum_{k=1}^{n} \bar{\lambda}i R(\vec{x}i, \omega)$$

This approach is very useful for carrying out gradex calculations within the drainage basin. The Gradex method [GUI 67] assumes that the gradex of the rainfall recorded in the drainage basin is equal to the gradex of the rainfall that is recorded in the local area. By using the two equations, it becomes very easy to calculate the amount of rainfall that has fallen in a drainage basin on a daily basis, and it then becomes possible to work out the gradex from these values. Figure 7.14 illustrates how this method can be adopted by using the example of the Ardèche drainage basin [BER 05]. The simplicity of the calculations makes it possible to calculate the exact amount of rainfall recorded in the basin on a daily basis by combining the values recorded in the drainage basin with daily rainfall values that are recorded by the network of rain gauges. Table 7.3 clearly shows that the average gradex of daily rainfall that is recorded by the rain gauges is at least 20% greater than the average gradex of daily rainfall that is recorded in the drainage basin.

This result clearly shows the advantage of using geostatistic tools when it comes to estimating rainfall levels in the drainage basin. The traditional methods used tend to produce systematic errors that cannot be corrected.

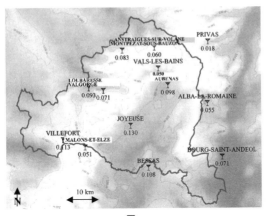

Figure 7.14. *The influence that $\overline{\lambda}i$ has on different rain gauges used to calculate extreme daily rainfall in the Ardèche drainage basin*

Drainage basin	Gradex average	Gradex of rainfall recorded in basin
ARVE	12.8	11.5
FIER	13.8	12.1
AIN	13.9	12.4
SAONE	10.8	7.6
ISERE	14.7	10.2
EYRIEUX	35.1	29.7
DROME	17.2	12.9
ARDECHE	38.6	29.1
CEZE	35.3	23.1
DURANCE	15.5	11.4
GARD	37.0	23.6

Table 7.3. *A comparison of the different estimations of the basin gradex values*

7.3.2.2. Precipitation is stationary

Sometimes, and this is often the case in urban hydrology, rainfall is said to be stationary (same average and same variance for each measurement site within the drainage basin) and that the spatial structure of the rainfall is said to be homogenous and isotropic (the correlation coefficient only depends on distance). If the term a_o is said to be zero, then the interpolation equation becomes:

$$\overset{*}{R}(\vec{x}o, \omega) = \sum_{i=1}^{n} \lambda i R(\vec{x}i, \omega)$$

This means that:

$$
[\rho_{ij}] =
\begin{vmatrix}
\rho(\vec{x}1, \vec{x}1) \rho(\vec{x}1, \vec{x}2)...\rho(\vec{x}1, \vec{x}j)...\rho(\vec{x}1, \vec{x}n) 1 \\
\rho(\vec{x}2, \vec{x}1) \rho(\vec{x}2, \vec{x}2)..\rho(\vec{x}2, \vec{x}j)..\rho(\vec{x}2, \vec{x}n) 1 \\
\rho(\vec{x}i, \vec{x}1) \rho(\vec{x}i, \vec{x}2)...\rho(\vec{x}i, \vec{x}j)...\rho(\vec{x}i, \vec{x}n) 1 \\
\rho(\vec{x}n, \vec{x}1) \rho(\vec{x}n, \vec{x}2)..\rho(\vec{x}n, \vec{x}j)..\rho(\vec{x}n, \vec{x}n) 1 \\
1 \quad\quad 1 \quad ... \quad 1 \quad ... \quad 1 \quad 0
\end{vmatrix}
\qquad
[\lambda_i] =
\begin{vmatrix}
\lambda 1 \\
\lambda 2 \\
\lambda i \\
\lambda n \\
\mu
\end{vmatrix}
\qquad
[\rho_{io}] =
\begin{vmatrix}
\rho(\vec{x}1, \vec{x}o) \\
\rho(\vec{x}2, \vec{x}o) \\
\rho(\vec{x}i, \vec{x}o) \\
\rho(\vec{x}n, \vec{x}o) \\
1
\end{vmatrix}
$$

The parameters λ_i are calculated by:

$$
[\lambda_i] = [\rho_{ij}]^{-1} [\rho_{io}]
$$

Imagine that there is a drainage basin made up of five rain gauges, as is the case in Figure 7.15; we assume that the correlogram takes the following form:

$$
\rho_{ij} = e^{-h_{ij}/P} = e^{-h_{ij}/8}
$$

The values of λ_{i_k} have been calculated for each point of the grid and from these values it is possible to create detailed maps similar to the one that can be seen in Figure 7.15.

From Figure 7.15, it can be seen that the isovalue curves are focused around the rain gauges and that the distribution of the grid points, which are part of the Thiessen polygons, are also coherent.

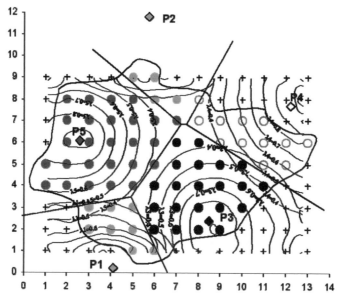

Figure 7.15. *The influence of λ_i on different stationary rain gauges*

	$\bar{\lambda}_1$	$\bar{\lambda}_2$	$\bar{\lambda}_3$	$\bar{\lambda}_4$	$\bar{\lambda}_5$
1	0.19	0.17	0.22	0.18	0.23
2	0.17	0.14	0.25	0.17	0.27
4	0.14	0.11	0.28	0.17	0.30
8	**0.13**	**0.10**	**0.20**	**0.16**	**0.32**
16	0.12	0.10	0.30	0.16	0.32
32	0.12	0.09	0.30	0.16	0.33
64	0.12	0.09	0.30	0.16	0.33
128	0.12	0.09	0.30	0.16	0.33
Thiessen	0.11	0.06	0.30	0.17	0.37

Table 7.4. *The evolution of $\bar{\lambda}i$ with range p*

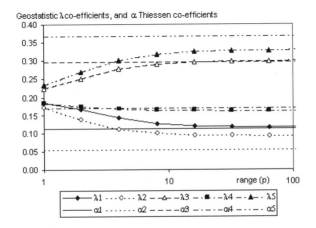

Geostatistic λ co-efficients, and α Thiessen co-efficients

Figure 7.16. *The evolution of $\bar{\lambda}i$ with an increase in the range and the comparison of $\bar{\lambda}i$ with Thiessen coefficients*

The averages of the values calculated by λ_{i_k} for different points on the grid and for different points within the drainage basin lead to the creation of weighting coefficients $\bar{\lambda}i$ from the equation:

$$R(\omega)_{BV} = \bar{a}o + \sum_{k=1}^{n} \bar{\lambda}i R(\bar{x}i, \omega)$$

With this equation it is clear that the values of the weighting coefficients change according to the spatial structure of rainfall. The calculations of $\bar{\lambda}i$ have been used once again, although this time the range p has been increased from 1 to 128. Table 7.4 shows the digital values of $\bar{\lambda}i$ and Figure 7.16 shows the increase in the range of p from 1 to 128. The first observation is that if the range p is very small then the value of $\bar{\lambda}i$ tends to fluctuate towards 0.2. This means that the rainfall recorded in the drainage basin is equivalent to the arithmetic average of the rainfall recorded by the rain gauges. However, when the range p is more structured (an increasing value of p) the value of $\overline{\lambda}i$ tends to fluctuate towards more stable values. These more stable values are quite similar to Thiessen's weighting coefficients, however, there is still a large enough difference between the two sets of values so that a geostatistic calculation of $\overline{\lambda}i$ is possible.

7.3.2.3. *An evaluation of the rainfall recorded in the drainage basin*

There is no simple method that can be used to record the amount of rainfall that has fallen in a drainage basin. If the linear interpolations of local rainfall are valid, the estimation of a particular level of rainfall in a drainage basin is equivalent to a linear combination of such observations. However, weighting coefficients depend on the spatial structure of rainfall, on the location of the measurement sites, on the contours of the drainage basin, and on the average values of local rainfall. Only methods generated from the methods that were introduced much earlier in this chapter can be used to give objective estimations as averages; this is made possible by using the geographical information that is available.

7.4. Conclusion

For the foreseeable future hydrologists will not have access to the most relevant information they require to find a solution to the problems associated with the estimation of rainfall levels. In order to overcome the challenges, hydrologists need to use data that provides abundant information on the flow of water at each measurement site in a drainage basin. Climate data (especially data relating to rainfall) is the only source of information that can be used to create the variable of temporal distance. This information can also be used to generate statistical methods used to estimate rainfall levels for a particular area. The information, which is produced by the networks of rain gauges, remains local because the methods used to estimate the amount of rain that falls in a given area (weather radars and remote sensing) are often imprecise and are limited to small areas. The information that is provided by the rain gauges relates to a recent period of time; for example the information that is available rarely dates back to a period of more than 10 years. The spatial interpolation of rainfall will remain a major concern for hydrologists for the next few years to come.

The networks of rain gauges that have been developed to record rainfall levels are not always adapted to the needs of hydrologists. The number of networks created that are used for research purposes depends on the length of time that the hydrologists will carry out their research. For example, more networks of rain gauges will be developed for a longer period of research so that more observations can be made. The following assumption can be made: the spatial range of rainfall varies just as the duration of rainfall varies but at a rate that is 0.4 times more powerful. This means that if a network is used to estimate daily rainfall levels, its density (which is measured in km^2) needs to be multiplied by 13 so that hourly rainfalls can be calculated. Geostatistic analysis makes it possible to compare the spatial structure of the rainfall with the density of the network that is required. As far as the networks of rain gauges are concerned it is possible to create different networks with different densities [CHA 00].

The information that is produced by the rain gauges provides information on temporal distance, but does not always provide accurate information on the spatial distribution of rainfall. For the last 25 years we have been working towards the following goal: it is necessary to include geographical information within the interpolation process that is applied to rainfall. Even if this geographical information were to be progressively adopted when it comes to plotting the parameters of precipitation on a map (median averages, values recorded over a period of 10 years, gradex, etc), this would not be the case for plotting the isohyetal lines on a map that provides information on a particular period of rainfall. Studies carried out in the French regions of Corsica and Provence-Alpes-Côte d'Azur (PACA) and in Algeria have shown that whenever geographical information is used in the interpolation process that is applied to rainfall, more information is known about the drainage basin. Using geographical information in conjunction with the interpolation process is only applied to certain parameters of hydrological models that are specific to the drainage basins. It is no longer used to correct values of the rainfall.

Initially, the geostatistical approach used to study rainfall seems to be rather complex and to have little to do with the physical make up of the rainfall. The geostatistical approach is, however, a particularly constructive approach in the sense that all hypotheses need to be clearly formulated, all uncertainties need to be quantified and at the end a coherent spatio-temporal model should be created.

7.5. Bibliography

[ALE 96] ALEXANDRE C., "Conception et dimenssionnement des barrages en terre de Mauritanie", *Mémoire d'ingénieur,* Ecole nationale du génie de l'eau et de l'environnement de Strasbourg, 1996.

[ASS 00] ASSABA M., LABORDE J.P., "L'intérêt de la prise en compte du relief dans l'estimation des pluies mensuelles: le cas de la Corse", *International Assocation for Climatology Publications*, vol. 13, pp. 149-157, 2000.

[ASS 04] ASSABA M., "La connaissance des pluies mensuelles au service de la modélisation hydrologique des apports mensuels en eau de surface", PhD thesis, University of Nice - Sophia Antipolis, 2004.

[ASS 06] ASSABA M., LABORDE J.P., ACHOUR F., "Global and distributed modelling of runoff in northern Algeria", 7th *International Conference on Hydroinformatics*, Nice, 2006, pp. 1551-1558.

[BEN 87] BENICHOU P. , LE BRETON O., "Prise en compte de la topographie pour la cartographie des champs pluviométriques statistiques", *Météorologie*, vol. 19, pp. 23-44, 1987.

[BER 00] BEROLO W., CHAMOUX C., LABORDE J.P., "Cartographie des précipitations annuelles, mensuelles et journalières extrêmes sur les Alpes Maritimes", *Publications de l'Association Internationale de Climatologie,* vol. 13, pp. 158-168, 2000.

[BER 03] BEROLO W., LABORDE J.P., *Carte au 1/200 000 des statistique des précipitations journalières extrêmes sur les Alpes Maritimes,* Conseil Général des Alpes maritimes and UNSA, Nice, 2003.

[BER 05] BEROLO W., LABORDE J.P., "Assessment of basin gradex over 1 to 7 days from daily rainfall data in the Rhone catchment", *6th International Conference on Hydroinformatics,* Seoul, 2005.

[BER 87] BERGAOUI M., DESBORDES M., "Etude de la structure spatio-temporelle des pluies à des échelles fines de temps et d'espace", *Research into Hydrology, held by the ORSTOM,* Montpellier, 1987, pp. 240-250.

[BRU 66] BRUNET-MORET Y., ROCHE M., "Etude théorique et méthodologique de l'abattement des pluies", *Cahiers ORSTOM série Hydrologie,* vol. 4, pp. 3-13, 1966.

[CAR 94] CARREGA P., "Topoclimatologie et habitat", PhD thesis, University of Nice - Sophia Antipolis, 1994.

[CHA 00] CHAMOUX C., LABORDE J.P., "Approche méthodologique pour la constitution d'un plan d'échantillonnage des pluies mensuelles et annuelles dans le département des Alpes-Maritimes", *International Association for Climatology Publications,* vol. 13, pp. 433-443, 2000.

[CHE 03] CHEIKHO T., "Synthèse spatio-temporelle des paramètres hydroclimatiques et modélisation hydrologique: application au bassin versant du Var", PhD thesis, University of Nice - Sophia Antipolis, 2003.

[CRE 79] CREUTIN J.D., "Méthodes d'interpolation optimale des champs hydrométéorologiques - comparaison et application à une série d'épisodes cévenols", PhD thesis, Grenoble Institute of Technology, 1979.

[DEL 78] DELHOMME J. P., "Kriging in the hydrosciences", *Adv. Water Resour.,* vol. 1, pp. 251–266, 1978.

[DEP 91] DEPRAETERE C., *DEMIURGE 2,0 une chaîne de production et de traitement de modèles numériques de terrain: TOPOLOG, OROLOG, LAMONT,* ORSTOM Editions, Montpellier, 1991.

[DJE 01] DJERBOUA A., "Prédétermination des pluies et crues extrêmes dans les Alpes franco-italiennes", PhD thesis, Grenoble Institute of Technology, 2001.

[GAL 82] GALEA G., MICHEL C., OBERLIN G, "Pluies de bassins: abattement sur une surface des averses de 1 h à 24 h", *Antony: CEMAGREF, (Antony Think tank),* no. 54, 1982.

[GOU 05] GOURBESVILLE P., LABORDE J.P., "Incertitudes et interrogations dans l'évaluation de l'aléa en milieu urbain: mesures, concepts et modèles", *La Houille Blanche,* vol. 1, pp. 60-64, 2005.

[GUI 67] GUILLOT P., DUBAND D., "La méthode du GRADEX pour le calcul de la probabilité des crues rares à partir des pluies", *AISH Publications,* no. 84, pp. 560-569, 1967.

[HUB 84] HUBERT P., *Eaupuscule: une introduction à la gestion de l'eau*, Ellipses, Paris, 1984.

[JOR 86] JORDAN J.P., MEYLAN P., "Estimation spatiale des précipitations dans l'ouest de la Suisse par la méthode du krigeage", *IAS*, vol. 13, pp. 187-189, 1986.

[JOU 80] JOURNEL A., HULJBREGTS C., *Mining Geostatistics*, Academic Press, New York, 1980.

[KIE 01] KIEFFER WEISS A., BOIS P., "Topographic effects on statistical characteristics of heavy rainfall and mapping in the French Alps", *J Appl. Meteorol.*, vol. 40, no. 4, pp. 720–740, 2001.

[LAB 02] LABORDE J.P., TRABOULSI M., "Cartographie automatique des précipitations: application aux précipitations moyenne annuelles du Moyen-Orient", *International Association for Climatology Publications*, vol. 14, pp. 296-303, 2002.

[LAB 03] LABORDE J.P., ASSABA M., BELHOULI L., "Les chroniques mensuelles de pluies de bassin : un préalable à l'étude des écoulements en Algérie", *Int. Conf on Water-risk Management in Semi-arid Countries,* Tunis, 2003, pp. 41-50.

[LAB 76] LABORDE J.P., "Notion d'indice de pente: approche par le calcul automatique", *Sciences de la Terre, Série Informatique Géologique*, n° 84, 1976.

[LAB 80] LABORDE J.P. , ZUMSTEIN F., THERRIOT D., "Mise en évidence des relations entre le gradex des pluies en 24 heures et les gradex des pluies de durées inférieures, *La Météorologie*", vol. 20-21, pp. 139-149, 1980.

[LAB 82] LABORDE J.P., "Cartographie automatique des caractéristiques pluviométriques: exemple de prise en compte de la morphométrie", *La Houille Blanche*, vol. 4, pp. 330-338, 1982.

[LAB 84] LABORDE J.P., "Analyse des données et cartographie automatique en hydrologie: éléments d'hydrologie lorraine", PhD thesis, National Polytechnic Institute of Lorraine, 1984.

[LAB 86] LABORDE J.P., "Pour une approche géostatistique des coefficients d'abattement", *La Houille Blanche*, vol. 3, pp. 221-228, 1986.

[LAB 95] LABORDE J.P., "Les différentes étapes d'une cartographie automatique: exemple de la carte pluviométrique de l'Algérie du Nord", *International Assocation for Climatology Publications*, vol. 8, pp. 37-46, 1995.

[LAV 97] LAVABRE J., CAMBON, J.P., FOLTON, C., MAKHLOUF, Z. ,MICHEL C., "LOIEAU: un logiciel pour l'estimation régionale de la ressource en eau : application à la détermination des débits de référence de la région méditerranéenne française", *Ingénieries – EAT,* vol. 12, pp. 59- 66, 1997.

[LEB 84] LEBEL T, "Moyenne spatiale de la pluie sur un bassin versant: estimation optimale, génération stochastique et gradex des valeurs extrêmes", PhD thesis, Joseph Fourier University, Grenoble, 1984.

[LEB 88] LEBEL T., LABORDE J.P., "A geostatistical approach for areal rainfall statistics assesment", *Stoch. Hydrol. Hydraulics*, vol. 2, pp. 245-261, 1988.

[LEM 86] LEMPEREUR R., LABORDE J.P., "Les pluies sur de petits bassins versants: une fonction aléatoire dont on peut estimer le variogramme", *Hydrol. Continentale*, vol. 1, n°. 1, pp. 3-13, 1986.

[MAL 74] MALLET J.L., "Présentation d'un ensemble de méthodes et techniques de la cartographie automatique numérique", *Sciences de la Terre, Série Informatique Géologique*, n° 4, 1974.

[MAT 71] MATHERON G., "La théorie des variables régionalisées et ses applications", *Les cahiers du CMM de Fontainebleau, Fasc. 5, Ecole des Mines de Paris*, Fasc. 5, pp. 1-212, 1971.

[OBL 87] OBLED C., "Introduction au krigeage à l'usage des hydrologues", *Research into Hydrology, held by ORSTOM*, Montpellier, pp. 174-222, 1987.

[SCH 77] SCHERER J.C., "Une méthode d'extrapolation dans l'espace de données pluviométriques moyennes", *Recherche Géographique à Strasbourg*, vol. 4, pp. 69-85, 1977.

[TOU 81] TOURASSE P, "Analyses spatiales et temporelles de précipitations et utilisation opérationnelle dans un système de prévision des crues - application aux régions cévenoles", PhD thesis, Grenoble Institute of Technology, 1981.

[ZAH 00] ZAHAR Y., LABORDE J.P., "Les précipitations journalières extrêmes de Tunisie: Gradex et valeurs exceptionnelles", *International Association for Climatology Publications*, vol. 13, pp. 181-190, 2000.

Chapter 8

Geographical Information, Climatology and Forest Fires

8.1. Forest fires: associated risks and individual components

8.1.1. *Analysis of the climatic risks and constraints*

Climate factors tend to be indirect sources rather than direct sources, of risks and constraints that can be fatal for people. There is almost only one climatic factor capable of threatening human life in a short period of time and that is the cold. The effects of the cold are made even worse by the wind, which increases the loss of calories experienced by warm-blooded beings, in other words the idea of wind chill, which is an indication of the cold that is felt by the human body [BES 74].

The number of people who die from the cold each year is very low in comparison to the number of people who are indirect victims of the different climatic elements. An individual rain drop or an individual snowflake has never killed anyone (however, there are cases of deaths caused by large hailstones strikes). However, heavy rainfall leads to deaths associated with landslides and floods. Other ways in which people can be affected by the indirect effects of these climatic factors include fallen trees or other objects that have fallen to the ground, which have been caused by strong winds from tornados or tropical cyclones, or by winter storms coming from west in the temperate world.

This notion of the indirect effects that the different weather elements can have on human life can also be applied to avalanches or to forest fires, but for different reasons. When it comes to identifying the different risk components that exist, it is best to choose three different components, instead of the two traditional components that are normally chosen, in other words the hazard and the vulnerability of the area concerned [CAR 03]. Whenever natural risks are studied we often hear or read

Chapter written by Pierre CARREGA.

about the confusion that exists between the hazard and what impacts this hazard can have on human life. For example, flash floods or landslides are not hazards, but are the result of a combination of a hazard (in this case rainfall) and the susceptibility (a potential) of the area in which the landslide can take place. If the inhabitant of a villa or a company manager arrive at their respective properties and find that water levels have reached their front door, their main worry is to try and find the best solution to decrease the water level, or flood level, which is perfectly normal. The water level, or flood level, is referred to here as the hazard. This vision cannot be the vision of scientists or territorial managers, who have to understand the reasons why such a problem has occurred in the first place, and to try and find a way so that the problem does not occur again, or so that it occurs much less frequently without leading to too many social and economic problems. Therefore, the hazard is the source of the problem, the triggering factor of an event, which may lead to dangerous symptoms.

Forest fires, bush fires, or savannah fires only start and develop if certain conditions take place at the same time: the symptoms that develop (the flames) are always created in the same way, according to the same steps, and in identifying these steps, it is then possible to isolate the different components of the risks that are linked to the creation of a fire. For a flame to be created there needs to be a spark, however, other factors need to be present for a forest fire to develop. One of the most important factors is the current weather when the spark is ignited. Forest regions in wet areas do not experience many fires, on the opposite. Mediterranean forests or shrub land are both the cause and consequence of forest fires, and this is due to the summer droughts that occur in this region. Thus, a strong relationship undoubtedly exists between forest fires and climate. To understand this relationship, as well as its application (which are used to both understand and forecast such fires), it is necessary to obtain geographical information.

8.1.2. *The components of forest fires*

8.1.2.1. *Hazard*

As is the case for other natural risks, forest fires are indirectly linked to local climate factors or to local weather. However, the different climate factors, as well as the weather, dictate the way in which forest fires develop.

A forest fire (or fire of any type of vegetation) can be divided into three main parts, which tend to occur in chronological order. Initially, a flame or spark is required so that a forest fire can develop. The source of the flame is natural (from a flash of lightening), or as is the case in the Mediterranean region of France, the source of the flame tends to be linked to humans, started either accidently or maliciously. Statistics taken from the Prométhée file[1] [PRO 07] show that the cause of approximately 50% of all fires is unknown. This figure is much higher in

---

1 Prométhée is an e-data base platform that exists in France.

comparison to a country such as Spain, where huge efforts have been made to try and find out the exact origins of the fires so that preventative methods can be taken to fight against such fires. Only a small number of the remaining 50% of identified cases were created by arsonists, but it is not sure the behaviour of arsonists is the same, according to the weather conditions are dangerous or not.

Whether a fire is started on purpose or not, there needs to be a flame or a spark present. However, flames and sparks only develop into full-scale fires if the different conditions that are necessary for a fire to start are all present. Factors that lead to the creation of a forest fire include: the disposal of burning cigarette ends on the ground; uncontrolled burning; sparks that come from machines on building sites (for example, the fire in the Adrets in the department of the Var in the south of France burned 400 hectares on July 16, 2007); the overheating of car engines, which transfers the forest fire into the neighboring forest (the Mandelieu forest fire destroyed almost 1,500 hectares on the July 5, 2007), etc. The existence of a combustible (as far as forest fires are concerned the combustible is vegetation), and the presence of an environment prone to combustion need to be present. In other words the flame needs to have to a source that is flammable.

Even with the presence of a combustible, the conditions that influence inflammation and combustion need to be sufficient so that a fire can develop: this is not always possible whenever it rains or due to the dew that falls on the ground in the evening. It is at this point when the surrounding weather conditions influence what happens, not only current weather conditions but also previous weather conditions, which govern the water status of the ground.

8.1.2.2. *Susceptibility*

Vegetal biomass plays the role as a combustible and can burn rather intensely. The exact heat given off by the biomass varies, depending on these different parameters as follows:

– the nature and the quantity of vegetable oils (terpenes) and hydrocarbons present, both of which burn very easily in their gaseous states. The morphology and make-up of the vegetation whose surface:volume ratio favors a rapid combustion means that it is easier for smaller objects to burst into flames. For example, it is more difficult for a log to burst into flames than hay. Flames will spread much quicker and it is much easier for them to spread if an area is densely covered with vegetation and if all of the individual layers of vegetation can be easily reached by the flames, for example, moss, herbaceous areas, shrubs, and forest areas. This is why the notion of clearing areas around houses that are threatened by summer forest fires in the Mediterranean region of France, was made compulsory after the introduction of the 50 m (sometimes 100 m) law. These laws state that land within a 50 or 100 m radius of houses in the Mediterranean region of France must be cleared so that the risk of a forest fire starting is reduced. The herbaceous layer is completely removed and the shrubland layer is strongly reduced, and in some cases the forest layer is also reduced. If we take the land-use type of this region into consideration, as far as vegetation cover is concerned, and which can be represented as a raster in a geographic information system (GIS), it becomes clear that this type of geographical information is a very useful tool so that the spread capacity of the flames can be estimated. The process involves examining and measuring how sensitive an area is to fires which might burn through it [CAR 05];

– the water present in leaves and branches also reduces the effects of combustion. During combustion all of the water present in leaves is evaporated, and this evaporation means that energy is taken from the leaves (for example, 537 calories are removed per gram of water). This removal of energy means that this energy can no longer be transformed into heat, and as a result this decrease in calorific energy means that it becomes more difficult for a fire to spread and the speed at which it spreads decreases.

Parameters, other than the morphology and the nature of the combustible, influence the way in which a fire spreads, and also provide information the "susceptibility" of a particular area to fires. Topography is another parameter that should be taken into consideration and it acts at two different levels: first, topography influences wind direction, and all of the different weather factors that exist, especially in terms of exposure. Second, topography can change the angle at which the flame comes into contact with the surface of the ground. As far as a perfectly vertical flame is concerned (in other words there is no wind present), if the topographic slope is steep, it means that the angle is weak and as a result the energy of the flame is better transferred to vegetation, which in turn makes it easier for a fire to spread, as well as speeding up the rate at which it progresses. If this angle is increasingly closed, there is an increase in the amount of radiation (that is emitted by the flame), which affects the surface of the ground, and thus, the vegetation. In addition, there is an increased chance that the convection column will affect the tops of the plants. Wind also has the same effect, and as there tends to be wind on the slopes of mountains it can be assumed that slopes, and relief in general, tend to be areas that are affected by fires. Moreover, the wind is seen as being less a hazard factor (due to a minor role for fire ignition) than a key factor due to its major role in helping fires spread from one area to another.

When we consider all of the different factors that influence the way in which a fire develops, it is necessary to make a distinction between the different time periods that are used so that it becomes possible to explain the situation occurring at a particular moment:

– time scales linked to the relief transformations (slopes, etc.) during thousands of years and this time scale may be considered as being invariable;

– time scales related to human actions that may have been taking place for tens or hundreds of years, as the evolution of the vegetation cover; and also time scales of a few years or months, concerned by the creation of new houses, other buildings or roads, etc;

– time scales used to describe the state of the vegetation and the ground at different seasons, or weather types that has occurred in area over a period of several weeks, which influences the water content of the vegetation and of the ground;

– there is also the time scale that deals with all of the current weather factors and these can be measured in hours or minutes.

8.1.2.3. *Vulnerability*

The term vulnerability is used to describe all of the factors which group together the impacts that humans and their development have had on nature. The type of vegetation that is present in these areas plays a key role as far as the economic effects of fires are concerned. The problem with the Mediterranean forest is that it takes place in a complex topography, and it does not have any real economic value. This fact prevents a fire risk management as efficient as the one that exists in the department of Landes in the South West of France, which produces a best quality wood, which is very easy exploitable. The main problem associated with forest fires is that they become houses fires, and more, they pose a threat to human and animal life. Unfortunately, some forest fires reach the end of their course by arriving in towns or by burning around the perimeters of villages or towns. In the town of Hyères in the South of France, cars that were parked on a street near the forest were burnt on the August 1, 1989; and in July 1997, the southern French city of Marseille was surrounded by flames on all sides.

Certain prevention plans have been introduced in France with the aim of fighting against the building of houses inside of the forest, which are vulnerable to forest fires. These prevention plans include: plans to decrease the risk of forest fires, the publication of land-use maps and urban area maps. Although the amount of vegetation (which acts as a combustible) around the access roads that lead to these houses has been reduced; this decrease in vegetation is not enough to prevent a fire from spreading in the event of strong winds. In the event of a serious fire, the main priority of firefighters is to protect the lives of the inhabitants of these houses, as well as protecting the houses themselves, before fighting the flames through the forest.

The term vulnerability is also strongly linked to the strategic and tactical efforts adopted by mankind to decrease the susceptibility (such as introducing fire lines, clearing up areas in which there is vegetation, and creating less flammable trees species plantations). The vulnerability concerns also the efforts to prevent fire eclosions through public awareness campaigns, and to detect them early (through monitoring and air patrols) and methods that can be used to fight a fire, once it has been declared.

The use of geographical information in all of this is extremely important and it is being increasingly used within GIS. It is extremely important that the information available in each layer of the GIS be frequently updated in this high-risk domain, on the different inhabited areas, different roads, different types of vegetation cover that are present, or the different locations of the nearest fire stations, etc.

8.2. The influence that different climate and weather factors have on forest fires. The indexes

8.2.1. *The water content of vegetation determines how long it takes for the vegetation to catch fire and burn*

8.2.1.1. *Measuring the water content of plants*

The influence of weather factors on forest fires is important because they play a role in the combustion process by acting both on the fuel and as an oxidant to a fire, at different speed. In order to understand how a fire develops, or to forecast its development, we need able to evaluate these factors, and to measure them. This is important because these same weather factors will eventually be used in the framework of meteorological risk indexes.

In order to examine and measure the hydration levels of plants that are prone to burning, several different methods have been developed. These methods are based on either taking a direct measurement from the plants or by providing an estimation of the measurements.

The best method used to measure the water content of plants involves weighing the plant before and after the plant is sent to an incubator to make sure that all of the water present in plants cells has been completely evaporated. The water content of the plant is then calculated by using the following equation: water (humid weight – dry weight) / humid weight. This equation is used to create a dehydration rate, from which we can work out whether a plant will burn easily or not. If we want to obtain information in real time, it is not possible to use such a method because it requires too much time and too much effort.

In order to measure the water content of trees, a different process is used. The process is based on the micrometric expansion that occurs in the trunks, branches, and twigs. This micrometric expansion is caused due to the flow of water that takes place in these individual parts of the trees. Close monitoring of the flow of water makes it possible to work out the hydration rate of the trees, and as a result, it makes it possible to work out whether it is necessary to carry out a process of irrigation in the area. However, we are not aware of this process being used to protect forests from fires.

Two different methods have been developed to work out the dehydration rates of plants: one of these methods is recent and relies on the use of remote sensing, whereas the older method involves estimating the water reserves of the soil by calculating the soils' water budget.

– Remote sensing can be used to work out the emission and reflection capabilities of leaved surfaces. This is made possible by using different wave lengths or combinations of different wave lengths, which lead to the creation of indexes such as the Normalized Difference Vegetation Index (NDVI), which is the most common index that exists. The emission and reflection capabilities of leaved surfaces are linked to the chlorophyll that exists in the leaves. The photosynthesis activity is associated with the presence of water and for this chemical reaction to take place, water is needed:

$$CO_2 + H_2O + energy \rightarrow C_6H_{12}O_6$$

This method has not, however, become very widespread for several different reasons, including the fact that the orbiting satellites used to take the measurements do not pass very frequently, a clear sky is required so that the layers of vegetation on the ground can be seen, and the correlation that exists between the signal that is received by the satellite's sensor and the dehydration rate of the leaves is not very strong. As a matter of fact, other factors can also influence the dehydration rate of the leaves, such as insect attacks.

– If we now take a closer look at the water reserves of the soil the following can be noted: if we use the principle that there is a strong relationship that exists between the water content of the soil and the water content of the plant, the idea is then to evaluate water reserves, on a regional level in order to avoid any difficulties that might be linked to spatial variability. The soil is considered as being made up of one or several water reservoirs that are filled up by rainfall, which are emptied owing to the process of evapotranspiration. The term water reserve is used to refer to the plants useful water reserve, the one that belongs to the soils and the ground and does not refer to phreatic water.

If we take the situation of when the soil's water reserve is full (generally in winter time), it is possible to calculate how the water reserve is emptied by using a decreasing exponential function. This process is much more progressive compared with the one which occurs concerning the filling of soil water reserve by rainfalls. The water from rainfall is considered as being entirely seeped into the soil. The removal of run-off water is not validated for a specific given moment, however, different studies have shown that the water budget, which is calculated monthly or over a period of 10 days, is validated. The daily needs of people have influenced the results of the water budget, initially not scheduled for a daily time scale. But other factors also influence these results as the way in which potential evapotranspiration is calculated, or the spatial distribution of real soil water content. On an area of several kilometers, this approach leads to satisfactory results during the growing season, in other words from spring time to the beginning of autumn. From the month of September onwards the nights become a lot colder and, therefore, the temperature of the air becomes much closer to the dew point temperature, at least on the surface of the ground. The resulting condensation covers the grass and the other materials on the ground with dew, which, in turn, reduces the risk of a fire occurring, but only if the water has not been evaporated. In winter the relationship that exists between the soil and plants is weak due to the fact that the plants are in a dormant state and this means that they have a low water content. However, the real risk of a fire occurring is much greater than the water content of the soil suggests. This is because even though the soil may be close to its saturation point, the plants that are partially cut off from the ground in winter have an extremely low water content, and are, therefore, susceptible to fires.

In order to study the water reserve on a much finer scale, the only water reserve can be divided into several smaller reserves, with each reserve being located at different depth. In 1987, Carrega suggested working on both the more generalist traditional water reserve and on a more superficial reserve at the same time. The aim of this simultaneous research was to provide a better description of the water content of the first 10 cm of the soil. As a result, it would then be possible to give more importance to the influence that rainfall or even the dew has on the superficial water reserve, which translates as just a small increase in the level of the general water reserve [CAR 88].

8.2.1.2. *Geographical information and the water content of plants*

In order to work out the water content of plants, it is necessary to use geographical information, which provides information from a point or from a surface. If the method used to work out the water content involves the use of remote sensing then the use of geographical information is obvious. If the method used is based on an estimation of the rate at which a water reserve or several water reserves are filled, then geographical information is little used, except in the choice of the maximum water reserve (when the soil has become saturated). In the Mediterranean region of France, the value (expressed in millimeters) attributed to maximum soil water reserves is included between 200 mm (in other words 200 l/m^2) in plains and 50 mm in mountains with steep slopes. The value of 150 mm is the value that tends to be attributed most as this value can be easily adapted to variations in topography.

This means that the "average" topography of a particular region will determine the maximum soil water reserve value that is attributed to that region. The water reserves themselves change on a daily basis, according to a different number of physical mechanisms that rely on the use of a certain number of geographical variables. The same is true as far as measuring forest fires is concerned, where the geographical variables that influence them change on a day-to-day basis.

8.2.2. *Atmospheric conditions during a forest fire*

Atmospheric conditions strongly influence the way in which a forest fire spreads, and they influence both the combustion process and the state of the combustible. In general terms, any external supply of energy and dry air means the rate at which the flames progress increases. However, as these phenomena are complex and cannot be measured easily, they are represented and treated in different ways depending on who is measuring the two variables.

The contribution of energy mainly comes from the sun, and solar radiation is the main element that is measured. It is measured as power, in Watts per meter square, as being an instantaneous source of energy supply.

The radiation budget of the Earth's surface also takes other sources of energy into consideration, coming from both the Earth and the atmosphere. The radiation budget can be written as follows:

$$Rn = RG (1\text{-}a) - T + A$$

where:

– RG is the total solar radiation that is received by the Earth's surface. RG is the sum of the direct solar radiation and diffuse solar radiation that reaches the Earth's surface;

– a is the albedo, the reflection rate of sunlight which, in theory, ranges in value from 0 to 1, yet in nature the range of values is actually from 0.05 to 0.80;

– T is the amount of radiation that is emitted by the Earth (in infra-red, given the temperature of T);

– A is the amount of atmospheric radiation, which is also recorded in infra-red.

Therefore, it is possible to consider the amount of energy that is produced by the Earth, by either studying the value of T, or by studying the value of the Earth's surface ts, which is associated with the value of T thanks to the Stefan-Boltzmann law. According to this law, T is a partial function of ts[4]. It is quite easy to measure the value of ts using satellites, airplanes or even on the ground with the use of a simple radiometer. But the heterogenity of the Earth's surface increases as and when the area being studied increases.

Another way to consider the amount of energy emitted by the surface of the Earth is to use the temperature of the air ta. The relationship that exists between the variables of ta and ts is quite complex and depends strongly on how wet the surface of the ground is: the drier the surface of the ground becomes, the more the difference between ts and ta during the day increases. Statistics have shown that air temperature is a factor that plays a role in the development and progression of forest fires.

There is also another set of factors that plays an important role in the development of fires, and these factors are linked to air humidity, which can be divided into two subgroups: absolute and relative humidity.

– Absolute humidity (e) is the quantity of water vapor (an odorless and colorless gas) in the air, and can be referred to in many ways: vapor pressure (hPa), ratio of mass/volume (g/m^3), ratio of gasoline to air (g/kg of dry air), etc. Absolute humidity can also be expressed as a temperature: the dew point temperature td. The capacity of the air to contain water vapor decreases with temperature. If td represents the temperature of a mass of air that becomes saturated with water vapor and to which td is linked, then the value of td is then expressed as a quantity of saturated water vapor.

– Relative humidity (h) is a variable that is much easier to measure than absolute humidity (e), and can be measured by the use of a hair hygrometer, by a psychrometer, or by the use of an electronic sensor. Relative humidity explains the relationship that exists between absolute humidity and the maximum amount of water vapor (ew) that can be held in a mass of air for a specific temperature at a given moment in time. For example, the air temperature (ta) 30°C, and the dew point temperature of the air (td) is 18°C, this means that the absolute humidity of the air (e), also known as its vapor pressure is 20.6 hPa. With the temperature of ta being 30°C, the air is capable of containing more water vapor than the quantity e suggests: the value of ew is 42.4 hPa. The relative humidity is 48% and is worked out as follows (e/ew) x 100. If the air temperature decreases in the evening (in the same location, without a change in pressure) and if we assume that the absolute humidity (e) does not change, the following pattern occurs: the value of ta decreases, as does the value of ew, and if the value of ew reaches the same value of e then the ratio e/ew is equal to 1, and, therefore, relative humidity reaches a value of 100%. For example, the value of ta fell from 30°C to 18°C, in other words it fell to the same value of td. As a result condensation begins to appear because of the excess water vapor that was left over and which was a result of the fall in temperature. This consideration can be present in the form of dew or fog.

It seems that the value of h plays a more important role in the creation and spread of fires than the value of e does. During the winter, the values of e are low, if not extremely low, due to the low temperatures during this season and possibly due to the continental origin of this air. Even though the value of td is very low, the value of h can still remain quite high. For example in winter, an absolute humidity of 6.1 hPa can become saturated at a temperature of 0°C (h = 100%, ta and td are 0°C). In summer, however, a temperature of 30°C for ta, and a relative humidity of only 30%, means that the value of e is 12.7 hPa, in other words the value of the summer vapor pressure is more than two times higher than the absolute humidity recorded the previous winter.

It is indeed this lack of water in relation to how much water the air can contain (which depends on the air's temperature) that favors the combustion process. This phenomenon can also be represented by the saturation deficit ew – e, as well as being represented by relative humidity (h).

This lack of water can also be explained by the water deficit ETP-ETR, where ETP represents potential evapotranspiration and ETR represents actual evapotranspiration. ETR is the amount of water vapor released by a complex ground-vegetation system over a given period of time. It is generally not possible to estimate the amount of water vapor produced by such a system in a period less than a day, except in a laboratory.

ETP is a concept used to express the efficiency of a particular area for evaporating water. It is also used to indicate the exact amount of water a surface would evaporate should the provision of water not be limited. It can be compared to a type of virtual ETR in some ways. This potential of an area to evaporate water does not depend on the amount of water available but only on the set of factors that favor the evaporation process (temperature, radiation, humidity, and wind). The value of ETP can never be lower than the value of ETR. A low ETR value is, either due to an absence of water that is to be evaporated, irrespective of what the ETP value might be even very high like in the Sahara, or due to a low ETP, like in Spitzberg for instance.

The wind is the final major atmospheric variable that plays an important role in the combustion process. The wind can be measured using various different scales: meters per second; knots (kt), where 1,852 m/hour is the equivalent to 1 kt; in kilometers per hour; or on the Beaufort scale. The Beaufort scale tends to be used to measure wind speeds at sea. The wind tends to help fires more when the winds are stronger, at a maximum of 70 km/hour, which is a little less than 40 kt or a little less than 20 m/second. At speeds greater than this the wind forces the flame to lose a lot of its vertical height. This theoretical limit, however, should be considered with caution, because as soon as the relief of the landscape becomes rugged wind speeds start to differ and start to interfere with the flow of air. Wind is the source of oxygen for the fire, and also increases the chances of combustion taking place. This is caused by two different factors: first, by the surface of the radiant band, and second, by the constraints that the convection column is subjected to:

– the radiant band is the surface of the flame, which, when it is loses its vertical height because of the wind, extends outwards and burns vegetation that had not been burnt previously;

– the convection column, which is fueled by burning gases created by the combustion processes, is also leveled by the wind, which in turn sends gusts of fire towards the plants.

This information shows that the wind plays a major role in the development and spreading of fires, as we had expected.

8.2.3. *Meteorological risk factors associated with forest fires*

8.2.3.1. *The need for indexes*

The risk linked to a forest fire spreading sometimes through densely populated forest areas in the Mediterranean region of France has led to great efforts being undertaken to prevent any development of a new fire igniting. As soon as a fire is located, the aim of the emergency services is to put it out in the quickest time possible, because they are aware of the fact that flames will spread as time goes on at exponential speed, and as a result, it becomes more difficult to extinguish the fire. Obviously, the methods used to fight forest fires should be called upon as soon as possible after the alert of a fire burning has been given. The best way to achieve this is to establish a pre-alert phase in which the emergency services are informed of what may happen, and as a result they have time to prepare themselves and prepare the equipment that will be used to help fight such fires. The use of such a policy is not too expensive from both an economic and social viewpoint, if it is carried out effectively as far as risk management is concerned and if the correct methods to fight against the fire are used. A compromise needs to be reached so that the resources used are only used in the area where the risk of the fire spreading is extremely high so that resources are not wasted.

The different meteorological components that influence fires and their effects have been described earlier in this section. However, the weather that occurs at a place P and at a time T is the result of a combination of all of these different meteorological components. Each of the individual components can change at any given time and as a result be favorable or unfavorable to the development of a fire. For example, these components could lead to the production of cold and dry weather, or warm and humid weather, which is an example of two opposite weather conditions. However, on certain occasions the different components could group together and add their potential in a same direction. Whenever it rains during cold and wet periods, and when there is no wind, forests are in no danger of bursting into flames. On the opposite, periods of hot, dry, and windy weather in which there is strong solar radiation are the most dangerous weather conditions for forests, especially when plants have been dehydrated for several weeks without rainfall.

Certain types of atmospheric circulation favor the different combinations of these dangerous weather factors. An example of such an atmospheric condition is the Foehn wind, which is a hot, dry wind associated with an extremely clear and transparent sky, meaning that there is a high level of direct solar radiation. This type of wind tends to affect mountain slopes that are "down wind" (on the slope opposed to the incoming wind) [CAR 02]. For example, mountain slopes that are leeward may have a temperature range of 5-10°C higher than the temperature experienced on the windward slopes. The relative humidity of the leeward slopes can fall to below 20%, and sometimes to even below 10%, although this is not very common. From this information we can see that if this wind were to make contact with the ground, all of the variables necessary to start a fire would be present. These variables would also help a fire to spread rapidly. The majority of catastrophic fires that occur in the French Mediterranean region are started in this way.

Considering the costs involved in trying to prevent fires from spreading, the decision makers in charge of the operation find themselves faced with a daily dilemma: although such a very dangerous event does not occur all the time, nor are the risks associated with fires extremely low all the time. This is why, several decades ago, the idea of allocating a single value to fire risk was adopted. This approach involves using indexes that analyze the fire risk, by combining different climate variables, which can either work together or cancel each other out regarding their effects on the fire (as described in the previous section). This method is used instead of trying to assess individually each t meteorological variable known as being associated with fire risk.

In order for an index to work effectively, it should reduce the amount of false alarms and missed alerts that occur. A false alarm is much less serious than a fire alert that has been missed. However, too many false alarms means that the emergency services may begin to question whether an alert is genuine each time one is raised or they may even withdraw their services because of the waste of resources that is generated each time a false alert is raised. A missed fire alert (for example, a serious fire has occurred even though the meteorological risk index was low) must be avoided at all costs.

8.2.3.2. *What is included in the indexes?*

The aim of this section is not to give a list of all of the indexes that exist all over the world, but to show that they have one common goal: to provide help so that the most appropriate decision can be made to help fight a fire, or prevent it to start. A meteorological index does not say that a fire is going to occur, all the more so as what a fire will become is depending on "susceptibility" of the environment, such as forest or shrubland (see earlier in this chapter). Should a fire be declared, the aim of the index is to provide information on the easiness of fire to start, or how it will spread. The index can sometimes be used to provide information on both of ignition and spread, and is referred to as a mixed index. The index is used as an indicator to describe any difficulties that firefighters and forest workers might have to face in order to fight a fire, it also lets these service teams know what precautionary measures need to be taken before they set out to fight the fires.

Since the 1970s, in France, the Orieux index has been used, which was named after its main inventor. The Orieux index combines wind speed and the level of water present in the soil water reserve [ORI 79].

In 1985 Carrega developed an index that combined the dryness of the air with the variables of wind speed and the level of water present in the soil's water reserves [CAR 85; CAR 91], and then in another index developed [CAR 88], the variable of air temperature as well as a second soil water reserve, was added, although this time the water reserve was superficial (saturated at 10 mm). The reasons for including these two new variables were to elucidate the layer of water present in the litter and in the first few centimeters of the soil (as a result rainfall or by the dew point). The level of water found at this upper soil level can influence whether a fire will develop or not.

Drouet worked on the speed at which flames spread [DRO 88], whereas Sol based his work on the research of Carrega and Drouet and integrated it into a new operational index [SOL 90], which was then replaced by a more complex index, the

Canadian index. Several different indexes were developed in the world, in the 1990s. All of these different indexes have one point in common and that is that they incorporate the same physical phenomena into the indexes, but the physical phenomena are expressed differently. The different phenomena used include: energy, temperature, the presence of water in the air (relative humidity, the water deficit, saturation deficit etc), wind, and the water stress that is experienced by vegetation. What makes the indexes different from one another is the way in which their different variables are combined: they can be added together or multiplied, which sometimes has a strong impact on the results that are produced.

In the 1990s, European Community research programs were developed so that research into fires could be carried out, and as a result lots of progress was made in this field. Some of the research programs focused on the meteorological risk indexes during which the indexes were tested and compared with one another by recalculating the indexes after a devastating fire. Regarding the Minerve and MEGAFiReS research programs [VIE 94; BOV 98], they both concluded that the Canadian index produced the best results. The problem that has hampered the research in this field is the fact that the validation criteria used to test indexes are taken from real fires, and mainly based on the surface area of the fire. We know that the overall surface area of a fire depends on how violently the fire burns and also on many other factors that have nothing to do with climate, such as position and extension of the forest, methods used to fight a fire, and the time that has elapsed between the start of the fire, the alert being raised, and the arrival of help, etc.

The best method to test an index is to create authentically sized test fires on similar patches of land. However, this is not very easy to do, especially when it comes to trying to gather information for the dangerous values of the index, which is a dangerous process as it could lead to the creation of a fire that might go out of control, thus leading to the necessity to get many firemen and equipment to protect against the fires. This method is, nevertheless, the only method that can be used to validate an index without intervention of external factors which are not part of the specific domain that the index is supposed to represent.

8.3. Using geographical information to work out the meteorological risks associated with forest fires

8.3.1. *The spatial model of a risk: a methodological choice*

8.3.1.1. *Spatial stakes*

It is not always possible to verify the accuracy of a particular index, the major challenge that exists today is to find a solution that can be used to work out a way to represent the location of the risks, in other words to give a spatial representation of risk, that allows mapping.

The most frequent question posed by those who fight fires once a fire has been declared relates to the exact location of the fire and its surrounding areas. Different types of geographical information are necessary so that a solution to this problem can be found, such as information relating to the relief of the area, the land-use type, communication networks and what state they are in at the moment of the fire. All of this different information, which is necessary, means that if a GIS is used, each of the individual layers of the GIS needs to be updated constantly [JOL 02; CAR 07] so that, for example, a building site or any other event does not change the access routes to the incident. Meteorological information also needs to be included as part of geographical information and, in particular, information relating to the wind is extremely important as the wind is the major factor that influences the direction in which a fire will spread.

As far as the prevention of a fire is concerned, calculating the index which is to be used in a smaller area leads to better information being produced about a fire, and as a result it becomes easier to prevent a fire in a smaller area from progressing any further.

Obtaining information on a small spatial scale, which is the large scale for geographers (as scale is measured as a ratio, not as a mathematical area), is desirable or almost obligatory. It now becomes a question of studying the different methods that can be used to obtain this information. Irrespective of the scale studied, we are not able to create a real surface area, but rather to create a juxtaposition of points that can be found either at the nodes of a regular grid , or at the center of a primary surface (the pixel of the grid) that a point is supposed to represent, in this case.

8.3.1.2. *Methods used to recreate the information*

Two main methods are used to fill an area or, in more exact terms, to assign a value to each point, which is part of this grid. The first method includes a determinist approach, which comes from the world of physics and is based on the use of processes. The second method is an empirical approach, which is based on previous experiences that have been recorded, and uses a statistical approach. Although these two approaches are completely different, they do partially depend on one another.

Another issue is the question of time. In other words, what period of time will the spatial model be applied to. The two approaches that have just been introduced provide two different answers to this question.

8.3.1.2.1. The physical approach

Let us take the example of the creation of a map that shows wind speed on a small scale, and in which each point is separated from its nearest neighbor by a distance of approximately 100 m (the map could also be created to show air temperature 2 m above the ground's surface, or its relative humidity, etc). The aim is not to go into detail regarding the different phases used in the calculation of the index, but rather to use the laws of physics that explain how and why air moves. For example, the presence of an air pressure gradient is required, which could be provided by a type of model that works on a larger scale (smaller for Geographers!). In conjunction with a Digital Terrain Model (DTM) that has been created, a smaller model will take the information given by the larger model, and will give a wind map, taking into account different problems associated with relief into consideration. If the general pressure gradient (synoptic scale) is zero and if we want to work in a non hydrostatic atmosphere, the pressure gradient needs to be generated so that breezes that are experienced in reality can be produced. To generate this pressure gradient, in other words differences in local pressure, the meteorological model needs to be capable of producing differences in temperature on the surface of the ground, which means that the energy budget needs to be calculated for each point. So that the energy budget can be measured, the following variables also need to be measured: latent heat (evaporation-condensation), convection, conduction in the ground fluxes, and the radiation budget, including visible radiation (solar) and infra-red radiation (from the Earth and the atmosphere). The energy balance depends heavily on the vegetation that is growing on the ground (and on its color), and on its orientation in relation to the where the sun is (such as slope and exposure), etc. The main problem associated with such a task is the fact that the more information required on a more detailed level, the more the calculations and data storage increase at an unbelievable rate. If the resolution of a grid is doubled, this means that the time and effort required to work out the calculations are increased by a scale of four. Another important point often overshadowed by the calculations (which do not stop increasing), is the sheer amount of information that is required for the model to work efficiently. Generally, the information required is geographical information, including the type and characteristics of soils, vegetation and the roughness of the ground's surface. Climatic information is also required, such as the limpidity coefficient of the sky, for each point. Further information on this topic can be seen in Chapter 4 of this book.

The absence of information means that only estimations will be given and as a result, the precision of the model becomes increasingly weaker.

Other less complex solutions can also be used. However, they are less accurate and cannot be used under certain conditions; this is the case in some hydrostatic models that use linear equations [NAN 85; CAR 94]. The advantage of using these models is that they work at a quicker rate than some of the other models, owing to their simple principles: the relief of an area is considered as being an obstacle that the synoptic flow of air must overcome, given some constraints in terms of the thickness of the limit layer, the roughness of the surface and the vertical stability of the air, etc. (Figure 8.1).

The problem with such a model is that it is incapable of simulating any thermal breeze for any area at any given time: as a matter of fact, the thermal wind is due to the relief and to the topographic contrasts that exist and which must not be seen as disrupting the flow of air, but as generating it. Using such a model as a basis is possible; however, caution is required so that it becomes possible to be able to distinguish between a type of wind, which is referred to as a synoptic wind (on a general level), and a thermal breeze. Thermal breezes account for 80% of the winds that occur in sheltered from main synoptic fluxes Mediterranean areas, such as the in the region of Nice, France.

11/8/94, 50.00 * 260 deg, 7 m/s a 10 m, BVF=0.001,
CL= 1,000 m, BLMLT3D - sub-domain - grid size 200 m

Figure 8.1. *Example of a wind field recreated for the extreme Southeast of France by a non-linear, hydrostatic model (BLMLT3D). The different shades of grey correspond to different altitude levels. The grid is 200 m wide [CAR 94]*

As far as the climatic variables are considered, physical logic can be used to provide real-time information or to forecast. Physical logic is used as the basis of all weather forecasts, and is used very effectively for short-term and medium-term forecasts. Beyond a period of 8-15 days, the reliability of the forecasts decreases due to reasons explained by chaos theory (due to problems associated with initialization values and with the non-linearity of equations). Plotting a map with just the average values (and not a given moment) assumes that each average value has been calculated from every individual situations that have occurred, so that the determinist logic is respected.

Due to the difficulties associated with using this elegant and satisfactory method, from an intellectual point of view, we, as scientists tend to use the second method based on experience, especially if we are working out average values, quantiles (values that have a certain probability of being reached or exceeded) or even if the detailed scale at which we want to work exceeds the different scales that are used by the physical model.

8.3.1.2.2. The empirical approach

In theory, this approach is very different to the previous one, because rather than giving values to a variable by using determinist laws, the empirical approach is based on any measurement that is available so that any gaps that exist can be filled with a value. With this in mind, a large amount of data is required, which is similar to the information that is being researched (in our example this is wind). This means that a permanent or temporary network of measurements is also required so that searches can take place.

At this stage, there is now a choice to operate, between two different approaches: the first method is based on the use of spatial analysis (spatial interpolation), and the second method is based on the use of what can be referred to as environmental analysis. Each of these two methods has been described in more detail earlier in this book.

– Spatial analysis leads to the use of spatial autocorrelation between different points in one area. It relies on the use of Newton's laws of physics, and postulates the interdependence that exists between these different points is all the more greater so as the distance that exists between them is small. If we take a typical example, an unknown value of a particular point which is located half-way between two other known points (at a speed of 6 and 8 m/second) is equal to the arithmetic average of the values for these points (in other words 7 m/second). This equation is a non-linear equation, and this is why most automatic mapping software has a default setting where the equation is based on the inverse of the distance squared. Carrying out further studies in this field led to the creation of the field of geostatistics in the 1960s, which was initially used in the world of geology [MAT 63]. The best applications are the different types of kriging that exist nowadays.

Technically speaking, in order to create the unknown values of the points, it is necessary to use the values of the points that are known. This is only possible on the condition that the average area of influence (in other words the range) of each measuring site is larger than the distance that exists between each site. This is the main role of the variogram.

If the value of the range is satisfactory, meaning that wind speed can be recreated, it is recommended to check that no physical or logical factor prevents such an operation from taking place.

The diagram on the left in Figure 8.2 shows a particular point X, whose wind speed (or temperature, etc) we would like to work out by using the values of other neighboring points, whose values are known (points A to D).

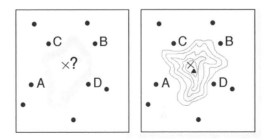

Figure 8.2. *The risks associated with spatial interpolation*

The diagram on the right shows the exact same information, except this time a major piece of geographical information has been added, relief. The relief is represented by the contour lines seen on the diagram. The relief of an area will dramatically change the values of the points. For example, altitude (as we can see, point X is at the top a mountain ridge) will produce values that are different from the values produced by the interpolation process that was carried out on the neighboring points in the box on the left where altitude was not a factor. This is because wind speed is probably much greater on a mountain ridge than it is on flat and low land. Temperature also varies, with the temperature that is recorded on the mountain ridge being much lower during the day in comparison to what the temperature on the flat land is.

If this method is used, it is assumed that there is a high level of homogenity that exists in the area we want to study for fear of producing any important errors.

– Environmental analysis works differently even if the space (distance) is integrated within the analysis. The basic idea is to search for laws that statistically link the value of a specific climate variable (wind, temperature, etc) to its surrounding area. The surrounding area needs to be represented in terms of quantitative data, and independent variables, which will be correlated to the dependent variable by using the process of linear regression. The first piece of information that tends to be used relates to altitude levels. There has been a lot of research into environmental analysis taking place since the 1970s, in which simple regressions in the form Y= ax+b have been used a lot.

It is thus possible to highlight a particular description variable that can be used to measure any preferential influence that a particular climate variable might have. However, the variations that exist between the different climate variables are multi-causal and additional information needs to be added to the model. One solution consists identifying clusters that gather all points responding to a particular criterion, which then considers all of the points as being homogenous. The regressions are then carried out inside of each cluster. The most obvious example is the one developed by. Douguédroit and de Saintignon [DOU 70]. Knowing that mountainous areas make it difficult to obtain high correlation coefficients between temperature and altitude, Douguédroit and de Saintignon started their research by separating the mountains areas into two sub-areas. On one hand they had the south-facing slopes, and on the other, they had the north-facing slopes and the bottoms of valleys. Each of the both sub-areas was subject to regressions, which improved the relationship that was being studied. The main difficulty using this method was linked to distinguishing boundaries, for example, when does a south-facing slope end and a north-facing slope begin? Where does the bottom of a valley begin? Another problem was linked to how the border and limits between two different areas could be managed. This is an excellent example of some of the problems that exist when we want to draw out geographical information.

In the 1980s [CAR 82; LAB 84; JOL 87], the development of multiple regression software makes possible to add descriptive geographical variables. In addition to altitude, other variables could be used, such as slope, exposure, position of the area in relation to a local level used as a basis, distance from the sea (if there is one), latitude, and longitude (if necessary), etc.

With multiple regression software a hyper-plan is used where the number of dimensions used is equal to the total number of variables that are used. Remember that the point cloud equation is written as:

$$Y = a_0 + a_1x_1 + a_2x_2 + \dots + a^nx^n + \varepsilon$$

whilst the regression equation that summarizes a cloud in a hyper-plan is written as:

$$Y' = a_0 + a_1x_1 + a_2x_2 + \dots + a_nx_n$$

Where ε represents the set of residuals, the sum of the squares of the distances that exist between the points of the cloud and the hyper-plan. The performance of the adjustment increases as the value of ε is weaker:

- Y = the observed value of the dependent variable,
- Y' = the value that is estimated by the equation,
- $x_1, x_2 \dots x_n$ are independent variables (regressors),
- $a_1, a_2 \dots an$ are regression coefficients,
- a_0 is a constant.

The multiple determination coefficient (R^2) refers to the part of the variance that is considered by the regression, with σ being the standard deviation for the series:

$$\sigma^2 Y = \sigma^2 Y. R^2 + \sigma^2 Y (1 - R^2)$$

with:

$$-\sigma^2 Y = \text{total variance of } Y,$$

$-\sigma^2$ Y. R^2 = the variance explained by the regression model,,

$-\sigma^2$ Y $(1-R^2)$ = variance of the residuals.

As is the case for a simple regression, each regression coefficient measures the change that takes place within the dependent variable Y, whenever an independent variable x_1, x_2..., x_n varies by one unit, the other variables remain constant.

In addition to the search for other variables, one of the keys behind the improvements that have been made to the model in case of a poor correlation, is the delinearization of the relationship that we want to obtain. The correlation can be improved by the use of a power law (for example, taking square roots), or by using a logarithmic law. The easiest way to improve the correlation is to keep the traditional calculation used to calculate the linear correlation coefficient , which was developed by Bravais-Pearson, and to transform the data by anamorphosis. As a result of the calculations carried out, a data matrix is created, in which, for instance, the variable "distance from the sea" does not relate to the total distance from the sea but rather to the square root of the total distance.

This approach, which is focused on geography, requires some previous knowledge on the topic, which is based on experience [CAR 94"]. The equation generated as a result is all the more representative of the statistical relationships that exist between climate and the environment so as correlation coefficient is high. Resolving the equation for each value of Y for which the independent X variables are known, makes it possible to provide data for the entire area in question.

At this point for each regression a choice needs to be made between the quest for the best regressors (independent variables), and to use a list that has been created containing a weak number of X variables which are always the same. It is also a very good idea to monitor the influence that altitude and distance from the sea have on these variables, in other words, to check whether the effects of altitude and distance from the sea are constant or whether they change on a daily, weekly, or monthly basis, etc.

The formal separation of the determinist and empirical approaches is not so clear cut: the choice of independent variables used is linked to the world of physics. Altitude decreases the effect that the ground roughness has on the wind, and it expresses a decrease in air pressure that influences air temperature. Distance from the sea influences the frequency and intensity of sea air advection currents. Relative altitude (which compares the location of an area to a base level) controls the nocturnal thermal inversion, which is linked to a strong level of terrestrial radiation at night-time etc.

Other possibilities also exist, such as those that are part of the NUATMOS model [ROS 88], which combines the advantages of both the empirical approach (using data generated by a network) and the advantages associated with fluid mechanics (because it takes the relief of an area into consideration). This model functions in two different phases: first of all it carries out a process of spatial interpolation, similar to kriging, by using data that has been captured by a network of sensors (it is possible to use different heights for each measurement site) without having to take the effects of altitude or relief into account. The second phase involves using the variables of relief and altitude as a basis for the model so that the first wind field that has been generated can be changed. The reason it is changed is to make it more realistic, due to the fact that the flow of air generated during the first phase is associated with the topography of the area [CAR 94; GUA 95]. In order to produce a wind field of very high quality, each major topographic entity (valley or mountain ridge) in which the research is taking place, needs to have a sensor present on it. This problem can be overcome by adding virtual sensors (generated by experts) to the sensors that already exist in the network [CAR 96].

For each of the meteorological and climate variables considered, it would seem that the use of the environmental analysis is better than the use of the spatial analysis that was described earlier in this section, as soon as the topography in an area is no longer flat and starts to get increasingly steeper. Some authors who have published works in this field have compared the performances between the two methods and their conclusions have confirmed what we had predicted [LHO 05]. However it is always satisfying to carry out research for a spatial structure within the regression residuals. The regression residuals are the differences that exist between the values that have been observed in a sample and the values that have been estimated by the regression model. If the variogram detects the existence of this spatial residual structure then it is recommended that a process of spatial interpolation be carried out on the residuals. The value of the residual calculated by the kriging process (or by any other spatial technique) should then be added to the value of each point in the area in which the research is taking place.

8.3.2. *A spatial model of the risk indexes*

8.3.2.1. *Area studied: the French Department of the Alpes-Maritimes*

The methods used in which geographical information plays a key role can be applied to many different domains. In this section we use the example of the spatial model that has been developed for the meteorological risk index that is applied to forest fires. The research was carried out along with Nuno Jeronimo.

Measuring the risk of a forest fire occurring is carried out on a daily basis by Météo-France in each of the predefined sectors. There are at least 100 different sectors that exist along France's Mediterranean coastline. The department of the Alpes-Maritimes is divided into seven of these sectors, with each sector having a different surface area (Figure 8.3).

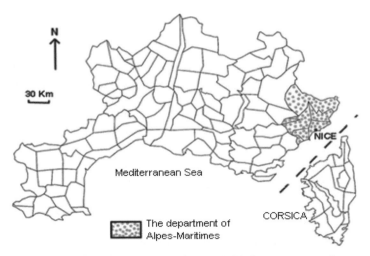

Figure 8.3. *The different sectors of France's Mediterranean coastline.*
The risk index that is used to measure forest fires, which is
calculated by Météo-France, is applied to these sectors

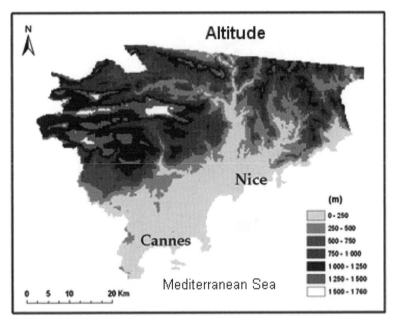

Figure 8.4. *The relief of the southern half of the department of the Alpes-Maritimes in the*
southeast of France (pixel = 50 m).Image created by ArcView and BD Alti

Geographically speaking, the Alpes-Maritimes is an area full of contrasts. For
example, its relief is very different, depending on whether we are talking about the
north or the south of the department. For example, the highest mountain peaks at

3,143 m and some of these mountains can be found very close to the sea. The northern part of the department is the most mountainous part and this is where the mountains reach their highest peaks, and as a result it is this half of the department that experiences less forest fires because due to many different climatic reasons. This is why the following example only deals with the southern half of the department (Figure 8.4).

The Digital Elevation Model (DEM) which was used as part of the research was the one that was created by the National Geographical Institute, known as BD Alti. The GIS that was used was ArcView, ArcGis 8.3 and it was used in raster mode. The use of ArcView made it possible to find out immediately what the exact altitude of a particular pixel was. Almost instantaneously, it generated values for the variable being researched, such as exposure, slope, relative altitude, and distance from the sea. Today GISs are equipped with integrated functions so that they can be used to perform immediate calculations of certain parameters, but these parameters do not necessarily correspond to the parameters that we want to work with (Figures 8.5 and 8.6).

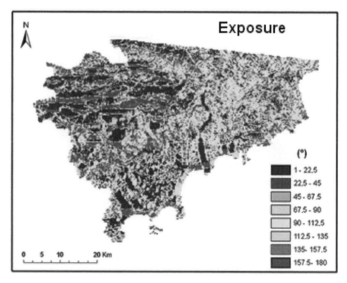

Figure 8.5. *A map showing the exposure of each pixel (from 1° in the north to 180° in the south; the east and the west had the same values)*

The slope of any given pixel is generally produced from the values of the altitudes of its neighboring pixels, which does not correspond to our research. Our research involves the search for a particular slope, which takes into consideration the average steepness within a 600 m radius. The value of this newly defined slope is then used in any future regressions, and is also used to provide information on the characteristics of each pixel.

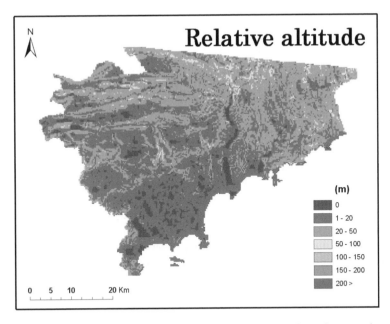

Figure 8.6. *A map showing the relative altitude of each pixel in relation to their surrounding pixels within a radius of 200 m (see color section)*

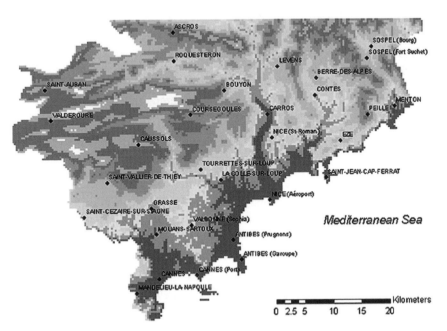

Figure 8.7. *A network map of the stations, and of the topographic context in which they were used*

The data produced that was used to constitute the series of regressions, was generated by automated stations that were initially financed by the Departmental Council of the Alpes-Maritimes. Twenty-three of these stations were selected and provided information relating to temperature, humidity, wind, and precipitation. The exact locations of the stations were known; therefore, it is possible to work out the exact geographical variables for each site (Figure 8.7).

8.3.2.2. The meteorological risk index "Carrega I85/90"

In this section we have decided to focus on the risk index that was first developed in 1985 and then modified in 1990 so that a scale ranging from 0-20 could be allocated to a specific risk [CAR 91]. Despite its simplicity, this index was recognized for its reliability and high-quality output and was partially used as the source for the Météo-France index (DK), developed by Drouet and Carrega at the end of the 1980s. The 1985 index, which was subsequently updated in 1990, was also used by firefighters and forest workers who worked in the Alpes-Maritimes area for more than 10 years as part of the system known as "Expert-graph", which was developed by the École des Mines in Paris [WYB 91]. The École des Mines is a prestigious French engineering school in Paris that has a campus based at Sophia Antipolis in Nice, which focuses on risk analysis.

The index's formula is written as follows:

$$I85/90 = (500 - (R^{0.5} * H / V)) / 25$$

where:

– R is the soil's water reserve, which is calculated by using a simplified Penman or Thornthwaite water budget. The aim of using such a water budget is to make the level of water present in the soil's water reserve similar to the level of water found in living plants (regional maximum level is fixed at 150 mm);

– H is the relative humidity;

– V is wind speed.

The value of the index varies from 0 to 20 (where a value of 20 represents maximum danger). The index values are divided into four main groups that are as follows: 0 to 8 represents no risk or a weak risk, 8-14 is used to represent an average risk, 14-18 is used to represent a severe risk, and 18-20 is used to represent a very severe risk. A value of 19 or greater is exceptional.

8.3.2.3. An important methodological choice needs to be made: do we recreate the different components of the index, or do we recreate the index itself?

The geographical information available, which has been measured by a GIS, can be used in two ways to create a map that represents the meteorological risk index [CAR 07]:

– The best approach to be used involves recreating the individual variables that make up the index. In the example of the Alpes-Maritimes, this would involve recreating the soil's water reserves, air humidity, and the wind. The water reserve evolves when it slowly begins to empty (except whenever there is rainfall present). Thus, the state of the water reserve on a given day D depends on the state of the water reserve on day D-1. Once the water reserve is full, it needs to be monitored on a daily basis. On the opposite, the wind and humidity on a given day D are completely independent of what they were like on day D-1, and this is because these two variables can change in a few minutes. In order to produce the definitive map, which shows the index for each point of the grid in an area, successive wind maps, reserve maps, and humidity maps need to be created. The index is then calculated for each point by combining all of the different maps that have been created.

– Another approach can also be used to recreate directly the index. This approach involves using the 23 measurement sites for which the index was initially calculated. In this approach, it is not necessary to recreate each individual component of the index.

8.3.3. *The results*

8.3.3.1. *The indirect recreation of the nocturnal meteorological risks, recorded at 03:00 UTC*

The most elegant solution involves recreating the spatial fields of the variables from which the risk index will be calculated.

Figure 8.8 is an example of the recreation of the relative humidity of the air by using the process of environmental regression and taking the following variables into account: altitude, relative altitude, slope, aspect, and distance from the sea (these variables are also used for the other examples of regression which follow in this section). A detailed examination means that it is possible to check how much influence the relief of an area and its morphology has on humidity, with a determination coefficient ($R2$) of 0.59. The partial correlation coefficients show that relative altitude has more influence than absolute altitude on nocturnal regressions. The least humid pixels can be found on the south-facing slopes, especially to the east of Nice, and areas that are quite far from the sea.

The wind, however, is very difficult to recreate for many different reasons including the fact that it varies strongly from area to area. And it has weak relationship and poor statistical correlation with descriptive environmental variables, such as those that are used to recreate humidity.

Figure 8.9 shows the recreation of the wind field when a different method, other than environmental regression is used. The method that was used in this case was spatial autocorrelation, including the process of universal kriging. As is to be expected, the effects of bumps and hollows are clear to be seen. It is also easy to locate several of the sensors because the influence that they have had on their surrounding area can also be seen. There is no justification as to why either of these two methods should be used. Other than using a digital model or a diagnostic model, such as Nuatmos, we are left with the choice of using the processes of regression or kriging as and when required.

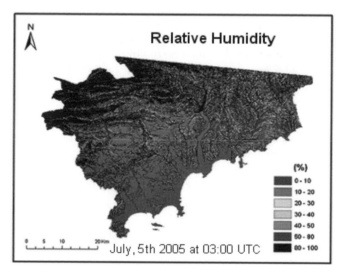

Figure 8.8. *Recreation of relative air humidity, by environmental regression on July 5, 2005 at 03:00 UTC*

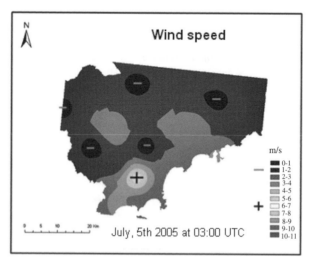

Figure 8.9. *Recreation of wind speed, by kriging, on July 5, 2005 at 03:00 UTC*

The soil's water reserve has also been recreated as part of our research carried out on July 5, 2005 (Figure 8.10). Regarding this variable, it is only possible to work on a daily level as it is only calculated once a day, in the evening of D-1 (after all of the water has been removed by the process of evapotranspiration during the day). The amount of water that has evaporated during the night was judged as being not important, and that explains the D morning reserve is equal to the D-1 evening reserve. The map highlights the low altitudes and the proximity to the sea as two factors that greatly influence the way in which the soil remains dry, because the summer rainfalls falling in the region are caused by the thunderstorms resulting from the convection currents that occur in the relief of the hinterland of the Nice area, far from the sea. In this example, the water reserve of the soil in the French Riviera only reaches a third of its maximum capacity of 150 mm. This is a real risk that could lead to the creation of a fire.

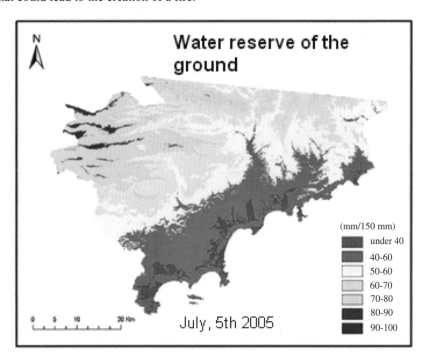

Figure 8.10. *Recreation of the ground's water reserve, by environmental regression, recorded in the early morning of July 5, 2005*

By combining the three maps shown in Figures 8.8-8.10 and by using the risk index equation, the result is the creation of the meteorological risk map, which can be seen in Figure 8.11. The risk of a fire occurring ranges from zero to average values in this case, depending on the area in question. This risk increases the closer towards the sea the area is, or in the southwest of the department (near the towns of Valbonne and Cannes) where it increases to severe owing to the stronger winds that blow in this area (Figure 8.9). By studying the map in closer detail, it is possible to see the two methods that were used to recreate the different variables:

– for example, the rounded marks that can be seen in Figure 8.9 are caused by spatial autocorrelation. These marks highlight the influence that the wind has on some of the measurement sites;

– in areas where the wind seems to be more constant, the slight risk differences are caused by the differences that exists in the area's relief.

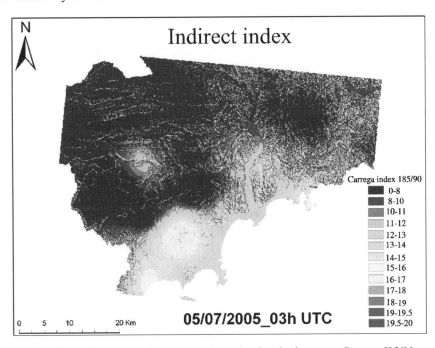

Figure 8.11. *Calculation of the meteorological risk index known as Carrega I85/90, during the night of July 5, 2005. Different successive layers were used, with each layer representing a different meteorological variable. The index is generated indirectly. Source: [CAR 07] (see color section)*

8.3.3.2. *The direct recreation of the nocturnal meteorological risk, recorded at 03:00 UTC*

Once the index for each measuring site has been calculated, it is very easy to recreate the nocturnal meteorological risk and to plot these values on a map. The determination coefficient of the process of environmental regression used reaches a value of 0.76 (or in other words 24% of variance, which is not "explained" by the model) and leads to the creation of Figure 8.12. The effects of bumps and hollows, which were caused by the interpolation method that was applied to the wind in order to recreate the index indirectly, are not present here. The only influences that can be observed here are the effects of relief and the proximity to the sea.

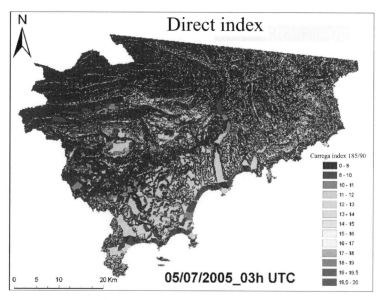

Figure 8.12. *Calculation of the meteorological risk index known as Carrega I85/90, during the night of July 5, 2005. The index is generated directly. Source: [CAR 07] (see color section)*

8.3.3.3. *The indirect recreation of the daytime meteorological risk, recorded at 15:00 UTC*

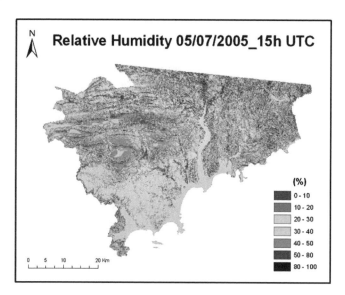

Figure 8.13. *Recreation of daytime relative humidity, by environmental progression, on July 5, 2005 at 15:00 UTC (see color section)*

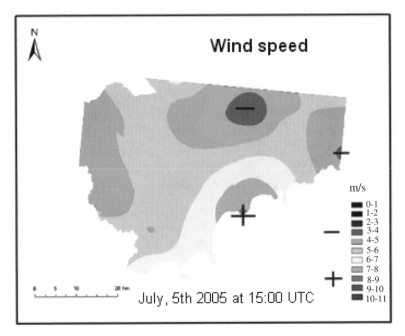

Figure 8.14. *Recreation of daytime wind speed, by using the process of kriging, on July 5, 2005 at 15:00 UTC*

Carrying out the same procedure during the day (data collected at 15:00 UTC) leads to the production of different results, because although the water reserve might be considered as being the same, relative humidity is lower during the day due to the higher temperatures (Figure 8.13). Relative humidity can be very low, reaching values of less than 30% along the French Riviera, these values can sometimes reach below 20%, whereas at night-time the values can range anywhere between 50 and 80%. The reason for the sharp decrease in relative humidity during the day is due to the fact that there is a southwest to west-southwest wind, which rises and is similar to the Foehn wind. This wind affects mainly the coastal area around Nice, and the areas to the west of Nice. The wind becomes stronger because during the day the cushion of stable air, which is established during the night and protects the ground from any synoptic winds, is not present (Figure 8.14).

The meteorological risks recreated during the daytime (at 15:00 UTC) are different to those created during the night because the effects of the bumps and hollows are less evident. This is due to the fact that wind gradients are more regular and homogeneous during the day, especially as far as the synoptic wind is concerned. This is not the case for the thermal breeze. It is possible to see the effect that wind has on the meteorological risk that reaches 19/20 in the windiest area, in other words on the coast (Figure 8.15).

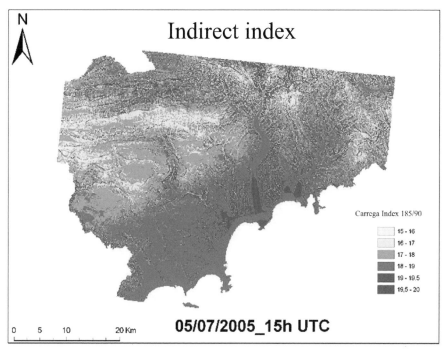

Figure 8.15. *Calculation of the meteorological risk index known as Carrega I85/90, on July 5, 2005 at 15:00 UTC. Different successive layers were used, with each layer representing a different meteorological variable. The index is generated indirectly. Source: [CAR 07] (see color section)*

8.3.3.4. The direct recreation of the daytime meteorological risk, recorded at 15:00 UTC

The direct recreation of the meteorological risk index, during the day or at night, leads to contrasting results, depending on the relief of the area. The direct recreation of the meteorological risk index leads to a risk value that could be in the range of 8-10 out of 20, and 19.5 out of 20. Whereas, the indirect recreation of the risk index generates a risk value of between 15 and 19.5 out of 20 (Figure 8.16). This shows that with the direct recreation of the risk index, there is a wider range of risk values used.

Contrary to indirect indice, for the direct indice the (synoptic) wind is not plainly interpolated, that increases the local effects associated with topography, and thus enhances the contrasts due to relief.

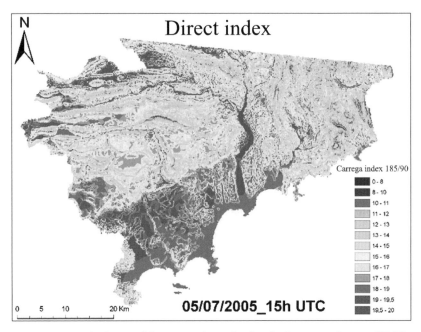

Figure 8.16. *Calculation of the meteorological risk index known as Carrega I85/90,*
on July 5, 2005 at 15:00 UTC.. The index is generated directly.
Source: [CAR 07] (see color section)

8.3.3.5. *A short summary of the two different methods*

Applying the process of kriging to the wind favors the creation of spatial areas whose gradient is not as steep as the gradient associated with the process of environmental regression. The process of environmental regression varies the risk index as far as the topography of the Earth's surface is concerned. This is an advantage with a thermal breeze regime in which the relief of an area play an important role. But, on July 5, the local effects of the wind gave way to a general flow, which led to the creation of a more homogenous spatial wind field due to the fact that large advection currents and the strong wind that is associated with such currents removed any local factors that might, otherwise, influence the wind.

If we take the example recorded on July 5, 2005 into consideration, the department of the Alpes-Maritimes in southeastern France could have experienced a potentially catastrophic forest fire due to the strong westerly Foehn winds (see earlier in this section), which started in the west of the region and became more widespread during the day with a hot, dry air, and a strong levels of solar radiation and high wind. This type of weather is particularly dangerous in summer (and more uncommon than in winter) , especially with the very low levels that water in the soil may reach (sometimes 5% of their maximum potential). However, there was more water present in the water reserves of the soil in the beginning of the summer in 2005.

On July 5, 2005 the extreme southeast of France was ravaged by large forest fires. For example, approximately 1,000 hectares of forest were burnt between the towns of Puget sur Argens and Fréjus (which are located in the east of the Var department), right next to the department of the Alpes-Maritimes. More than 100 hectares were burnt in the districts of Villeneuve-Loubet and Vallauris (Alpes-Maritimes), despite the fact that these areas are partially urbanized and that they have extremely efficient methods available for fighting such fires. The fire risk index of these areas was in the range of 17 and 19 out of 20, irrespective of whether the indirect or direct method was used to calculate the risk index, which highlights the fact that the method used to work out the fire risk index was extremely effective.

8.4. Conclusion

Throughout this chapter we have dealt with the topic of forest fires, which can be seen as being a risk that is partially associated with climate, and we can observe that geographical information is omnipresent throughout the chapter. Throughout this book, geographical information has been applied to different factors that have had an effect on the world of climatology, however, geographical information also influences other factors that, in turn, influence forest fires. Geographical information needs to be taken into consideration as far as other risk factors that are not directly associated with climatology are considered, like susceptibility or vulnerability. The aim of this is to be able to understand how the fires start and develop long-term strategies for better prevention and protection. In other words geographical information is used for operational purposes.

The aim of this chapter is not to use geographical information as a measurement that can be used in calculations, but to use it in as many different approaches as possible rather than only using it to recreate climatological data. This phase is not present in the direct method, which is used to recreate the risk index that is, nevertheless, a synthetic climate variable.

How valid is the map that represents information linked to a forest fire meteorological risk index, whatever the method used to elaborate it? This question needs to be asked at two levels: first, the index needs to be validated, yet not based on real fires. As we have seen in this chapter, forest fires are not dependent only on meteorology. An ideal index would not, therefore, be one that had the best correlation with burnt surfaces, or would not be the one that generates the least number of missed and false alerts. The best way to avoid missing any fire alerts is to constantly increase the risk level that has been set, as this leads to an unbelievable number of false alerts, and as a result of the high number of false alerts the emergency services may question the severity of any future risk levels. In theoretical terms, a good index should never lead to the creation of any missed alerts. However, if a false alert does occur there are two possible explanations as to why it might have occurred: first, the risk was exaggerated, and second, the index was not wrong, it was just the fact that there was no initial spark present to start the potentially catastrophic fire in the first place. We will never be sure if the absence of a fire, which was initially given a high risk index value, is due to the overestimation of the risk index and, therefore, due to the poor forecasting of the weather conditions for an area. Conversely, the weather conditions could have been forecast correctly and would, therefore, have led to the creation of a fire; however, no fire started because there was no initial flame or spark present to ignite it.

For an index to be validated there needs to a situation in which the hazard associated with lighting the fire is removed, and the only conditions that vary from one fire to another are the meteorological conditions that influence the fires and the way in which they develop. The solution to this issue is to ignite pieces of land that have been chosen for testing, with each plot having similar slopes and vegetation cover. The tests would be carried out during the most varied weather conditions. The idea is to correlate the fire behavior (speed of progression, height of flames, heat emitted, etc) with a particular index value. From there it would be clear that a violent fire cannot correspond to a low-risk index value. This approach of using experimental fires is carried out by our lab at the end of winter as part of a burning program developed by the French National Forestry Office in the department of the Alpes-Maritimes in the southeast of France. However, lots of these experiments need to be carried out in the most extreme and, therefore, most dangerous conditions so that all possibilities have been tested and can be protected and fought against should a real fire of such an extent occur in the future.

The next step involves validating the recreation of the spatial field of the risk index, by using any of the methods introduced throughout this chapter. The most common approach involves excluding a certain of number of points (whose index is calculated from meteorological data that has been recorded) from the experimental sample (for example, the file on which the regression is calculated). This approach also involves checking whether there are not too many residuals, with a residual being the difference that exists between the actual recorded value and the value that was predicted by the model.

Due to the fact that an index can be calculated directly but not measured directly, the difficulty associated with validating it is not easy to solve. At this point, validating the index in itself is more important than validating its spatialization.

In general terms, the use of geographical information is essential so that data relating to the dangers associated with fires can be produced irrespective of the method used to generate the information. Geographical information is used differently, depending on the method that is used to generate the data, and often the data are generated at different phases of the different methods. The risk is composed with hazard and susceptibility that must be associated to spatial areas, and risk is present only because of the relationship that exists between the vulnerability. As the vulnerability must also be associated to spatial areas, the use of geographical information is, therefore, required at all phases.

8.5. Bibliography

[BES 74] BESANCENOT J.P., "Premières données sur les stress bioclimatiques moyens en France", *Ann. Géogr.*, vol. 93, no. 459, pp. 497-530, 1974.

[BOV 98] BOVIO G., CAMIA A., "An analysis of large forest fires danger conditions in Europe", *IIIrd International Conference On Forest Fire Research, 14th Conference On Fire and Forest Meteorology*, vol. 1, Luso, November 16-20, 1998, pp. 975-994.

[CAR 82] CARREGA P., "Les facteurs climatiques limitants dans le sud des Alpes Occidentales", *Revue d'Analyse Spatiale Quantitvative et Appliquée*, n° 13, University of Nice, 1982.

[CAR 85] CARREGA P., "Une formule simple pour l'estimation du risque d'incendies de forêts dans les Alpes Maritimes", *Bulletin de la commission météorologique des Alpes-Maritimes*, Nice, 1984-1985.

[CAR 88] CARREGA P., "Une formule améliorée pour l'estimation des risques d'incendie de forêt dans les Alpes-Maritimes", *Revue Analyse Spatiale*, vol. 24, pp. 165-171, 1988.

[CAR 91] CARREGA P., "A Meteorological Index of Forest Fire Hazard in Mediterranean France", *Int. J. Wildland Fire*, vol. 1, no. 2, pp. 79-86, 1991.

[CAR 94] CARREGA P., GLINSKY- OLIVIER N., "Validation of wind models, Final report (2)", in *Modeling Forest Fires*, European contract EV5V-CT91-0015, Birkerod, June 1994.

[CAR 94"] CARREGA P., "A method for the reconstruction of mountain air temperatures with automatic catographic applications", *Theoretical and Applied Climatology*, vol. 52, n°1-2, pp.69-84.

[CAR 96] CARREGA P., "Une procédure d'identification automatique et en temps réel du régime de vent en cours, pour un système d'aide à la décision contre les incendies de forêt", *International Association for Climatology publications*, vol. 8, p. 13-21, 1996.

[CAR 02] CARREGA P., NAPOLI A., "Climat, Fœhns et incendies de forêts", *International Association for Climatology*, vol. 14, p. 35-43, 2002.

[CAR 03] CARREGA P., "Les risques naturels liés à la pluie et à la sécheresse: élaboration de cartes des pluies extrêmes et des risques d'incendies de forêt dans une région méditerranéenne: la Toscane (Italie)", in *Riscuri si Catastrofe*, vol. II, University Babes-Bolyai, Cluj-Napoca, Romania, pp. 271-286, 2003.

[CAR 05] CARREGA P., "Le risque d'incendie en forêt méditerranéenne semi-urbanisée: le feu de Cagnes–sur-mer (31 August 2003)", *L'espace Géogr.*, vol. 4, pp. 305-314, 2005.

[CAR 07] CARREGA P., JERONIMO N., "Risque météorologique d'incendie de forêt et méthodes de spatialisation pour une cartographie à fine échelle", in *Climat, tourisme, environnement, Proc. of the 20th Int. AIC Conf.*, Carthage, September 2007, pp. 168-173.

[DOU 70] DOUGUÉDROIT A., DE SAINTIGNON M.F., "Méthode d'étude de la décroissance des températures en montagne de latitude moyenne: exemples des Alpes françaises du Sud", *Rev. Géogr. Alpine*, pp. 453-472, 1970.

[DRO 88] DROUET J.-C., SOL B., "Etude de nouveaux indices de risques météorologiques d'incendies de forêt en zone méditerranéenne", I.U.T., Aix en Provence / C.I.R.C.O.S.C., Valabre, Service météorologique interrégional Sud-Est, Marseille-Marignane, 1988.

[GUA 95] GUARNIERI F., CARREGA P., GLISNKY-OLIVIER N., LARROUTUROU B., "Expert knowledge and quantitative wind modeling for spatial decision support dedicated to wildland fire prevention", *Proc. of the Int. Emergency Management and Engineering Conf.*, May 9-12, Nice, 1995, p. 171-176.

[JOL 87] JOLY D., "L'interpolation supervisée. Une méthode de traitement destinée à la cartographie automatique présentée à l'aide d'un exemple de climatologie", *Actes du Symposium Intern. sur la Topo-climatologie et ses applications*, Liège (Belgium), March 14-16, 1985, pp. 135-148.

[JOL 02] JOLY D., BROSSARD T., ELVEBAKK A., FURY R., NILSEN L., "Présentation d'un SIG pour l'interpolation de températures à grande échelle; application au piémont de deux glaciers (Spitsberg) ", *International Association for Climatology publications*, vol. 14, p. 287-295, 2002.

[LAB 84] LABORDE J.P., "Analyse des données et cartographie automatique en hydrologie. Eléments d'hydrologie lorraine", PhD thesis, Travaux Universitaires 1984.

[LAB 95] LABORDE J.P., "Les différentes étapes d'une cartographie automatique: exemple de la carte pluviométrique de l'Algérie du Nord", *International Association for Climatology publications*, vol. 8, pp. 37-46, 1995.

[LHO 05] LHOTELLIER R., "Spatialisation des températures en zone de montagne alpine, PhD thesis", University of Grenoble 1 - Joseph Fourier, 2005.

[MAT 63] MATHERON J.C., *Traité de géostatistique appliquée. Le krigeage*, vol. 2, Dissertations written by the B.R.G.M., n° 24, Paris, 1963.

[NAN 85] NANNI S.C., TAMPIERI F., "A linear investigation on separation in laminar and turbulent boundary layers over low hills and valleys", Nuovo Cimento, vol. 8C, pp. 579-601, 1985.

[ORI 79] ORIEUX A., "Conditions météorologiques et incendies de forêt en région méditerranéenne", note technique no. 8, section XXIV, Météorologie nationale, Paris, May 1979.

[PRO 07] FICHIER P., La banque de données sur les incendies de forêt en région méditerranéenne en France, 2009. Online at http://www.promethee.com/prom/home.do.

[ROS 88] ROSS D.G. *et al.*, *Diagnostic wind field modelling: development and validation*, NERDDP, project n° 1040, Center for Applied Mathematical Modelling, Chishlom Institute of Technology, Australia, 1988.

[SOL 90] SOL B., "Estimation du risque météorologique d'incendies de forêt dans le Sud-Est de la France", *Rev. forestière française*, pp. 263-271, 1990.

[VIE 94] VIEGAS D.X., SOL B., BOVIO G., *et al.*, Comparison study of various methods of fire danger evaluation in Southern Europe, in *Proceedings of 2nd Int. Conf. on Forest Fire Research*, vol. 2C-05, Coimbra, pp. 571-590, 1994.

[WYB 91] WYBO J.L., EXPERTGRAPH. "Analyse basée sur des connaissances et suivi en temps réel d'information géographique évolutive. Application à la prévention des incendies de forêt", PhD thesis, University of Nice-Sophia Antipolis, 1991.

List of Authors

Maria João ALCOFORADO
Institute of Geography and Territorial Planning (IGOT-UL)
University of Lisbon
Portugal

Pierre BESSEMOULIN
Météo-France and
World Meteorological Organization (WMO) Commission for Climatology (CCl)
France

Pierre CARREGA
Laboratory UMR ESPACE/CNRS
University of Nice-Sophia Antipolis
Nice
France

Vincent DUBREUIL
Laboratory COSTEL/CNRS
University of Rennes 2
France

Daniel JOLY
Laboratory THÉMA/CNRS
University of Franche-Comté
France

Jean-Pierre LABORDE
Laboratory UMR ESPACE/CNRS
University of Nice-Sophia Antipolis
Nice
France

Isabelle ROUSSEL
Professor Emeritus
University of Lille 1
France

Wolfgang SCHOENER
University of Vienna and
Department of Zentralanstalt für Meteorologie und Geodynamik
Vienna
Austria

Index